ASPIRE
SUCCEED
PROGRESS

Complete Physics for Cambridge Secondary 1

Helen Reynolds

WORKBOOK

Oxford excellence for Cambridge Secondary 1

OXFORD

OXFORD
UNIVERSITY PRESS

Great Clarendon Street, Oxford OX2 6DP

Oxford University Press is a department of the University of Oxford.
It furthers the University's objective of excellence in research, scholarship, and education by publishing worldwide in

Oxford New York

Auckland Cape Town Dar es Salaam Hong Kong Karachi
Kuala Lumpur Madrid Melbourne Mexico City Nairobi
New Delhi Shanghai Taipei Toronto

With offices in

Argentina Austria Brazil Chile Czech Republic France Greece
Guatemala Hungary Italy Japan Poland Portugal Singapore
South Korea Switzerland Thailand Turkey Ukraine Vietnam

First published in 2013

British Library Cataloguing in Publication Data

Data available

ISBN 978-0-19-839025-1

20 19 18 17 16 15 14 13

Printed in Great Britain by Bell and Bain Ltd. Glasgow

Acknowledgments

®IGCSE is the registered trademark of Cambridge International Examinations.

Cover photo: PASIEKA/Getty

Artwork by: Q2A Media and Erwin Haya.

Welcome to your **Complete Physics for Cambridge Secondary 1** Workbook.

This Workbook accompanies the Student Book and includes one page of questions for every two pages of the Student Book. Each question page includes several types of question.

- Some questions ask you to choose words to complete sentences. These questions will help you to learn and remember key facts about the topic.
- Other questions ask you to identify statements as true or false, or put statements in the correct order. Some of these questions are testing your knowledge, others are asking you to apply what you know to a new situation.
- There are many questions that ask you to interpret data from investigations, or information from other sources. When you answer these questions, you will be practising important science skills, as well as preparing for the Cambridge Checkpoint test.
- Some pages include comprehension questions. They ask you to read some information, and then answer questions about it. Many of these questions will help you develop skills of evaluation.
- Most pages have an extension box. Some of these questions will help you to extend and develop your science skills. Many others go beyond Cambridge Checkpoint Science. They include content from the Cambridge IGCSE® course. All the extension questions are designed to challenge you, and make you think hard. There aren't any spaces for your answers to these extension questions, so you'll need to work on a separate sheet of paper.

This Workbook has other features to help you succeed in Cambridge Checkpoint and eventually Cambridge IGCSE:

- The glossary explains the meanings of important science words. It includes all the **bold** words in the Student Book, and others.
- The Cambridge Checkpoint-style questions near the back of the book are excellent practice for the Cambridge Checkpoint test.
- The exam-style questions show you what you are aiming for. Give them a try!

I wish you every success in science, and hope you enjoy the Workbook.

Table of contents

Stage 9

1 Circle the correct word or phrase in each **bold** pair in the sentences below.

The force that pulls objects towards the surface of the Earth is **gravity / air resistance**. The forces that slow down objects that are moving on solid ground are **gravity / friction** and **electrostatic force / air resistance**. The forces of **air resistance / friction** and **water resistance / gravity** will act on anything falling through the air. A force called **water resistance / upthrust** acts on any object that is floating on water or submerged in water.

2 A student drew this diagram to show the forces acting on a boat that is floating on the water.

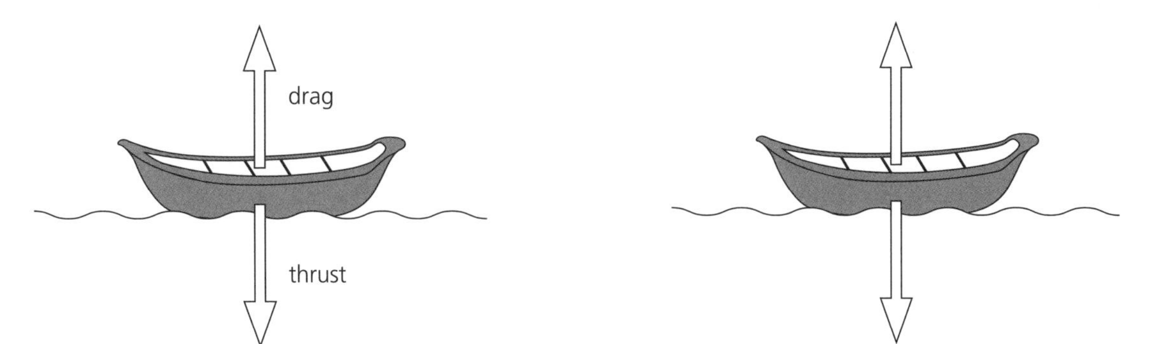

Explain what is wrong with the diagram. Label the arrows correctly on the second diagram.

....

....

3 A student measures the weight of a bag of bananas with a spring balance.

a Describe what is inside a spring balance and explain how it works.

....

....

b What is the name of the force that is acting *upwards* on the bag of bananas?

....

c The person selling the bananas in the market says that they have a 'weight' of 1 kg. Explain what is wrong with this statement.

....

....

E

Here is a list of forces:

weight	**electrostatic force**	**magnetic force**	**friction**	**air resistance**
	water resistance	**thrust**	**upthrust**	**tension**

a Which forces are contact forces, and which are non-contact forces? Explain your answer.

Forces can change the speed and the direction of motion of a moving object.

b Describe a situation in which water resistance changes the speed of a moving object.

c Describe a situation in which friction changes the direction of a moving object.

1 All of the sentences below are wrong. Circle **one** incorrect word in each sentence, then rewrite each one to make it correct.

a When a cyclist is decelerating uphill the forces on him are balanced.

....

b A cyclist cannot reach terminal velocity going downhill.

....

c When friction and upthrust are equal to thrust the cyclist is moving at a terminal velocity.

....

2 Here is a box with forces acting on it. Complete the sentences to describe what will happen to the box.

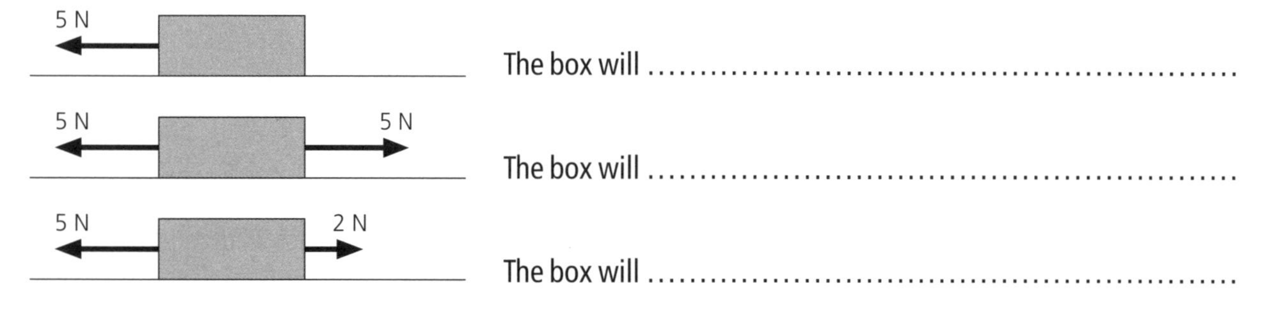

The box will ....

The box will ....

The box will ....

What instrument would you use to measure the forces on the block?

....

3 Write ***T*** next to the statements that are true. Write ***F*** next to the statements that are false. Then write corrected versions of the statements that are false.

a The forces on a floating object are balanced.

b The forces on a rocket taking off are balanced.

c The forces on a leaf accelerating to the ground are unbalanced.

d All objects that accelerate have balanced forces acting on them.

Corrected versions of false statements:

....

....

E

a Explain what is meant by a '**resultant**' force.

b Copy the table and complete the final column.

	Thrust of car (N)	Friction (N)	Air resistance (N)	Resultant force (N)
A	1500	700	800	
B	3000	500	100	
C	1000	600	500	
D	2500	900	400	

c The car is currently moving with a *steady speed.* What will happen to it in each case?

1 Circle the correct word or phrase in each **bold** pair in the sentences below.

The force of friction is measured in **newtons / kg** and acts between two surfaces that are **in contact / not in contact**. You can reduce the friction between two surfaces by using **streamlining / lubrication**. The direction that friction acts is always **the same as / opposite to** the direction of motion of the object.

2 In some situations it is helpful to have a large force of friction, but in others it is helpful if the force of friction is small. Write the words 'large' or 'small' in the final column.

Example	Large or small force of friction
The friction between your shoes and the ground when you walk should be …	
When you brake, the friction between your brake blocks and your bicycle wheel should be …	
The friction between the moving parts in a car engine should be …	
The friction between a tyre and the road should be …	

3 Match the beginning of each sentence with the correct end of the sentence. Connect the boxes together with lines.

Friction …
It is difficult to walk on an icy pavement because …
Matches light because …
Car brakes …

… of friction.
… rely on friction to work.
… there is not much friction.
… slows things down.

A student is investigating friction. Here is the apparatus that he uses. He puts different materials on the ramp and puts a block of wood on top. Each time he finds the height that he needs to lift the ramp for the block to start moving.

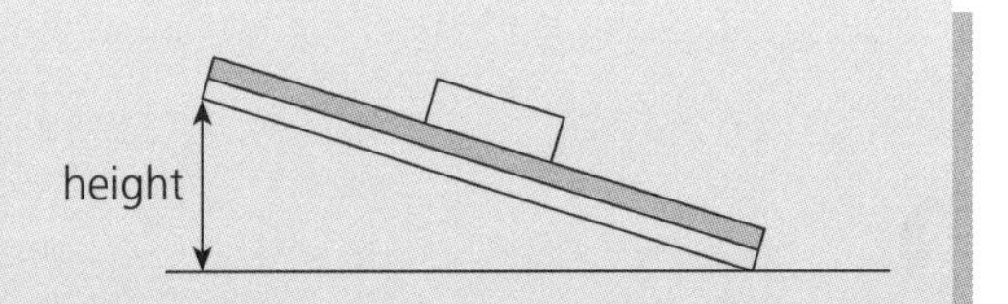

a Here are his results. Put the surfaces in order of least friction to most friction.

Surface	Height he needs to lift the ramp (cm)
wood	15
paper	17
carpet	25
sandpaper	20

b He tries the same experiment but this time with two blocks of wood, one on top of the other. What would happen to the height that he needs to lift the ramp now? Explain your answer.

1 Look at the diagram of the two people at two different points on the Earth. Draw arrows showing the force of gravity acting on each person.

2 A student is investigating the link between weight and mass. She takes some measurements and puts them in a table.

Mass (g)	Weight (.....)
100	1.0
150	1.5
200	3.0
250	2.5
300	3.0

a Fill in the unit of weight in the table.

b Which weight is incorrect? What is the correct weight?

....

3 The top drawing on the right shows part of a ride at a theme park.

a Which arrow shows the direction of the weight of the people sitting in the carriage? Write down the correct letter.

....

b The bottom drawing to the right shows a different ride where the carriages are upside down. Which arrow shows the direction of the people's weight while they are upside down? Write down the correct letter.

....

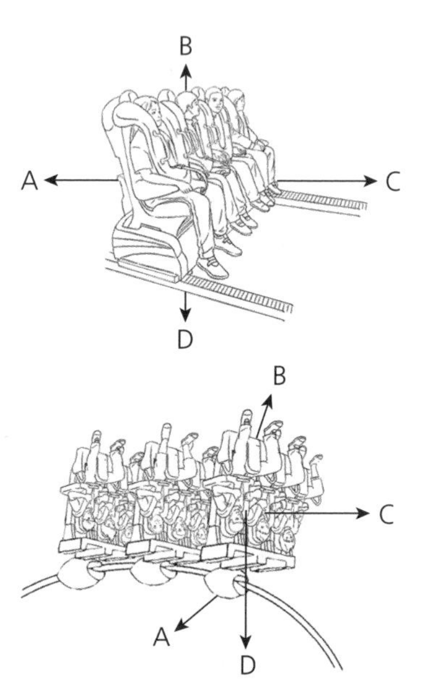

E

This table shows the mass and weight of objects on Earth, the Moon, and on a planet that is not in our Solar System. The gravitational field strength on Earth is 10 N/kg, and is 1.6 N/kg on the Moon.

Mass of object on Earth	Weight on Earth (N)	Mass of object on the Moon	Weight on the Moon (N)	Mass of object on the planet	Weight on the planet (N)
50 g					2.5
500 g					25
2 kg					100
25 kg					1250

a Copy and complete the table.

b Calculate the gravitational field strength on the planet.

c Is the mass of the planet more or less than the mass of the Earth? Explain your answer.

1.5 Questions, evidence, and explanations

1 Use the words and phrases from the box to complete the sentences below.
Use each word once, more than once, or not at all.

evidence	explanations	ideas	questions	data	observations	proof

Scientists ask .... about the world around them. They make ....

and use them to develop .... Some people think that they are looking for

.... that their .... are correct, but this is not the case. If the

.... supports their................................, then they are accepted.

2 Newton's idea about gravitation was very successful.

a Why did Newton think that there was a force acting on the Moon?

....

....

b Why was it hard for people to believe that the Earth was exerting a force on the Moon?

....

....

How scientists develop explanations	
Ask a question.	*Why is the orbit of Uranus not quite a circle?*
Suggest an explanation using scientific knowledge.	*There is another planet beyond it that is affecting the orbit because of gravity.*
Test the explanation by collecting data.	
Check the evidence. Does it support the explanation.	
The explanation is accepted.	

c Copy and complete the table to show how to use Newton's idea to make a prediction.

3 Write ***T*** next to the statements that are true. Write ***F*** next to the statements that are false. Then write corrected versions of the statements that are false.

a It is possible to prove that a scientific explanation is true.

b Newton's law of gravitation explains all observations.

c Newton's law of gravitation was used to send people to the Moon.

d If new evidence does not support an explanation, then scientists look for a new explanation.

Correct versions of false statements:

....

....

E

Bhaskaracharya's idea was that objects exert forces on each other, the force we now call gravity. When Bhaskaracharya told people about his ideas about gravity some people thought that the Earth must be falling down as well.

a Can the Earth be falling 'down' as well? Explain why this idea is not correct.

b How was Bhaskaracharya's idea about gravity similar to Newton's?

c How was it different?

1 Use the words and phrases from the box to complete the sentences below.
Use each word once, more than once, or not at all.

air resistance	balanced	constant	decreases	friction	faster	slower	unbalanced	streamlined

If the forward force on an object is equal to the forces of .... and ....,

we say that the forces are .... The forces of ....

and .... tend to slow moving objects down. You can reduce the force

of .... by making the object more ....

2 The sentences below describe the forces acting on a parachutist from the moment that she jumps out of the plane until she lands, but they are in the wrong order.

A When the air resistance is equal to the weight the parachutist will travel at a steady speed, called the terminal velocity.

B The parachutist is travelling slowly, so that she can land safely.

C The air resistance increases as she accelerates.

D The parachutist opens her parachute.

E The air resistance suddenly increases, so the parachutist suddenly slows down.

F The air resistance increases until it equals the weight of the parachutist and she reaches a lower terminal velocity than before.

G As the parachutist jumps out of the plane she accelerates.

Write the letters in the correct order.

3 Explain why:

a Birds take on a streamlined shape to dive into water and catch fish.

....

....

b Large parachutes slow you down more than small parachutes.

....

....

E

a Draw a diagram showing the forces acting on a piece of paper being dropped on Earth and on the Moon.

b Sketch a graph to show how the speed of the paper changes with time for each situation.

c An astronaut called David Scott did an experiment on the Moon in 1971. He dropped a hammer and a feather together and they hit the ground *at the same time.* Explain why.

1.7 Planning investigations

A student is planning an investigation into the strength of a bridge. She uses a model bridge made of cardboard. She puts masses on a piece of cardboard supported between two piles of books and measures the deflection.

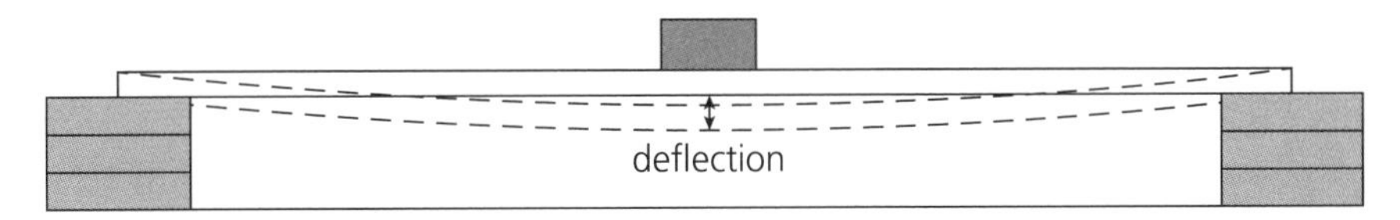

The student thinks of some questions to test. The first question that she thinks of is 'How does the distance between the books affect the deflection?'

1 Write down two other questions that she could answer in this investigation.

....

....

2 She decides to investigate how the distance between the books affects the deflection. What do you think will happen? Write your prediction below.

....

....

3 Write down all the variables in this investigation.

....

....

4 Make a list of all the equipment that the student needs.

....

....

5 Write a plan that shows how the student can do this investigation.

....

....

....

6 Draw a table for the results.

Describe one problem that the student might have doing this investigation. Describe how to deal with this problem.

1 Write ***T*** next to the statements that are true. Write ***F*** next to the statements that are false. Then write corrected versions of the statements that are false.

a An object in a liquid sinks when its weight is greater than the upthrust acting on it.

b An object floats when the upthrust is bigger than the weight.

c The force holding a climber on a climbing rope is upthrust.

d A spring that extends 2 cm with a force of 1 N will stretch 4 cm with a force of 2 N on it.

Corrected versions of false statements:

....

....

2 A company that organises bungee jumps has to choose a new bungee cord. They tested three different types of cord by adding weight to the cord until it snapped. They plotted their results on a graph. Use the information on the graph to answer the questions below.

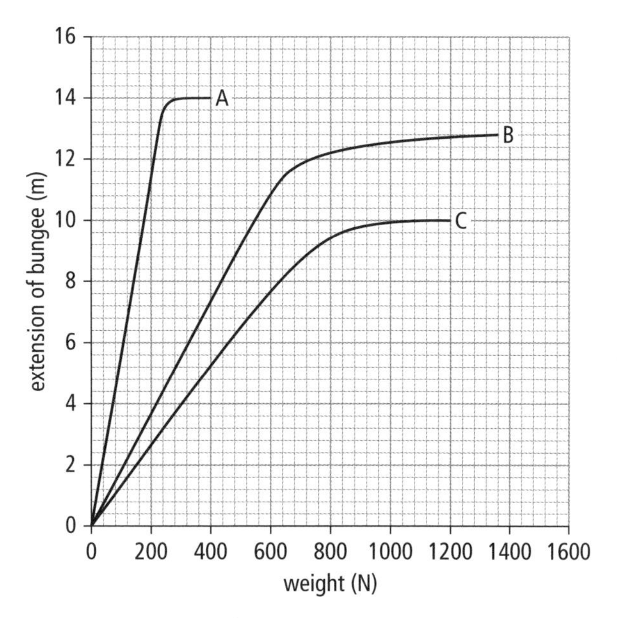

a Which cord requires the biggest force to break it? How big is that force?

....

b Which cord has the biggest extension? How big is that extension?

....

Five people want to try bungee jumping. The table shows some information about their weights.

Family member	Weight (N)
Aci	500
Bintang	1000
Citra	750
Dewi	1300
Elsa	380

c Which cord can *all* the people in the table use?

....

d Which cord can Bintang *not* use?

....

e Aci and Citra safely do a tandem jump (jump together) with one of the cords. Which one?

....

E

Two boys take a boat out onto a lake.

The lake is shallow at the sides and deeper in the middle.

The first boy says 'As we go into deeper water, the water level on the side of the boat will be higher.' The second boy says 'The water level on the side of the boat will not change.' Who do you agree with? Why?

1.9 Presenting results – tables and graphs

A student is doing an investigation into air resistance. He makes some parachutes out of paper and attaches them to a small mass. He drops each parachute and times how long it takes to reach the ground.

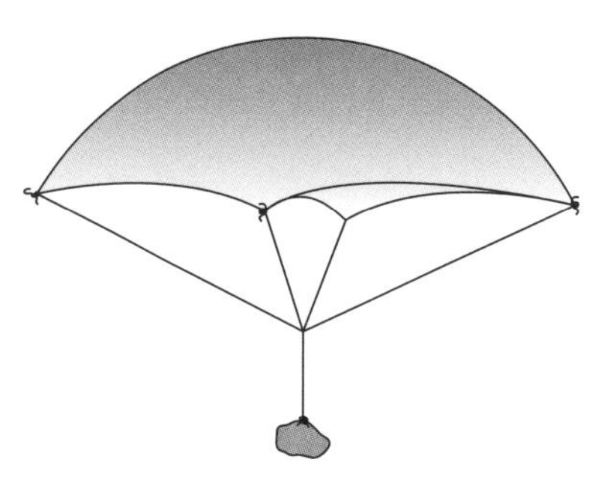

He makes some notes of the times that it takes different parachutes to reach the ground.

Area = 20 cm^2, time = 1.5 seconds.

Area = 40 cm^2, time = 2.2 seconds.

Area = 60 cm^2, time = 4.2 seconds.

Area = 80 cm^2, time = 3.6 seconds.

Area = 100 cm^2, time = 4.1 seconds.

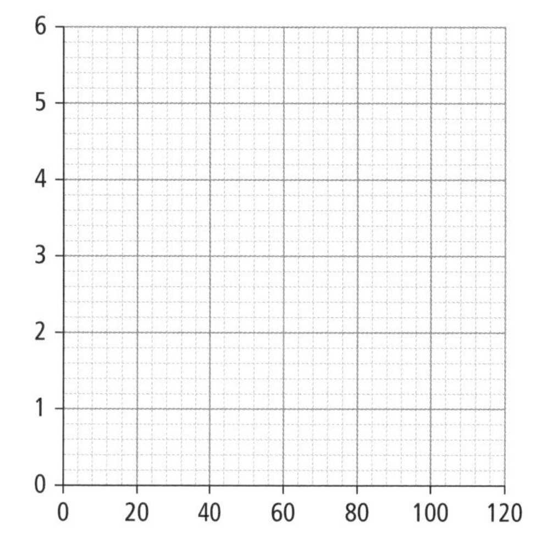

1 Draw a table of these results.

2 **a** Plot the results on the graph paper above. Draw a line of best fit.

b Explain why you have drawn the sort of graph that you have drawn.

....

c What is missing from the graph?

....

d Is there an anomalous point? Which point is it?

....

e What should the student do about this point?

....

E

a Does the line of best fit that you have drawn go through (0, 0)?

b If it does, explain why it does. If it does not, explain why it does not.

c Another student investigates how the type of material of the parachute affects the time that it takes to fall. How would the graph she draws be different from the graph that you have drawn above? Explain your answer.

1.10 Round in circles

1 Use the words and phrases from the box to complete the sentences below.
Use each word once, more than once, or not at all.

speed	direction	acceleration	centre	edge	centripetal	circular	tension	Earth	gravity

When an object moves in a circle its .... is constantly changing

even though the speed stays the same. There is a force acting on the object acting towards the

.... of the circle. This is called the ....

force. If you swing a ball on a string around your head, then this force is provided by .... The force of

.... between the .... and the Moon provides the .... force

needed to keep the Moon in a circular orbit.

2 There are lots of situations where things move in circles. This diagram shows a hammer thrower, at the point that he lets go of the hammer, looking from above. In which direction does the hammer move?

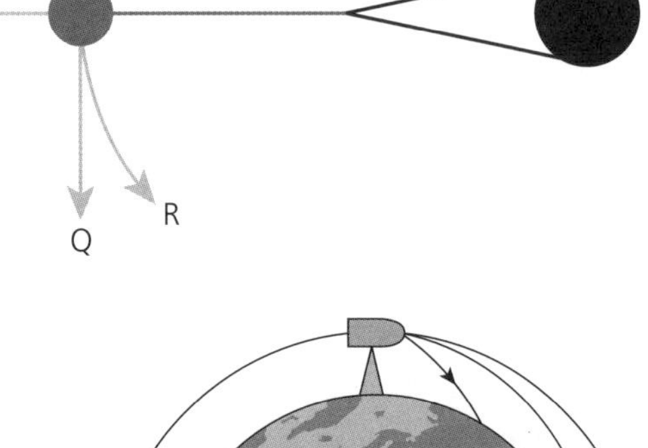

....

3 You can think about launching a satellite into orbit like this: imagine firing a cannon that is at the top of the mountain.

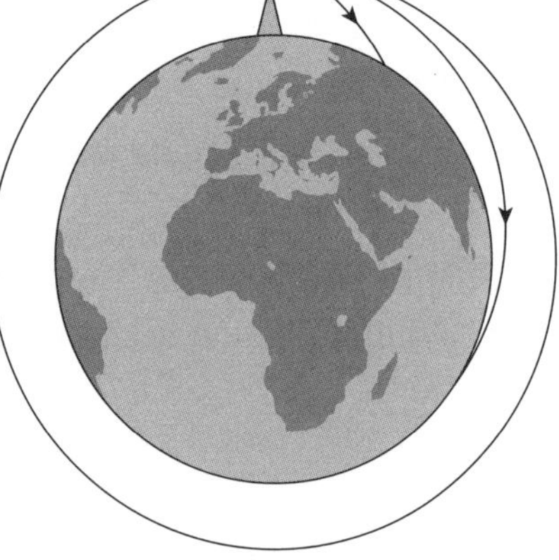

a What force pulls the cannonball to the ground?

....

b Explain how you can get the cannonball to go into orbit.

....

....

c A real satellite is launched upwards on a rocket. When it is at the correct height a small rocket provides a force that speeds up the satellite, just like the cannonball above. What would happen if the force on the satellite was too big?

....

....

4 This is a ride that you find at some fairgrounds.
Explain why the chains are designed to hold many times the weight of a person.

....

....

2.1 What is energy?

1 Use the words and phrases from the box to complete the sentences below.
Use each word once, more than once, or not at all.

joules	energy	0.001	fuels	food	money	newtons	1000

The .... that we need every day comes from the .... that we eat. Your body needs .... to keep warm, and to keep your body working. .... and .... are stores of energy. Energy is measured in .... or kilojoules. One kilojoule is ....

2 Here is the energy content of 1 g of some different fuels.

Fuel	Energy (kJ)
coal	20
oil	45
gas	40
wood	15

a Which fuel stores the most energy? ....

b Which fuel stores the least energy? ....

c How much energy in *joules* is there in 1 g of coal? ....

d What mass of wood has the same energy as 1 g of oil? ....

e Which fuel stores twice as much energy as coal? ....

3 Here is the energy needed for some activities.

Activity	Energy for each minute of activity (kJ)
sitting	6
cycling	25
football	59
swimming	73

a Why does your body need energy even when you are not moving?

....

b How many minutes would you need to cycle for to burn off 100 kJ?

....

c Why do children need more energy than they use for the activities that they do each day?

....

E

A student wants to measure the energy stored in some foods. He uses a small amount of each food to heat up a test tube of water. He measures the temperature of the water before and after he burns the food.

a State the variables that the student will need to control in this experiment.

b Predict the link between the energy content of the food and the temperature increase.

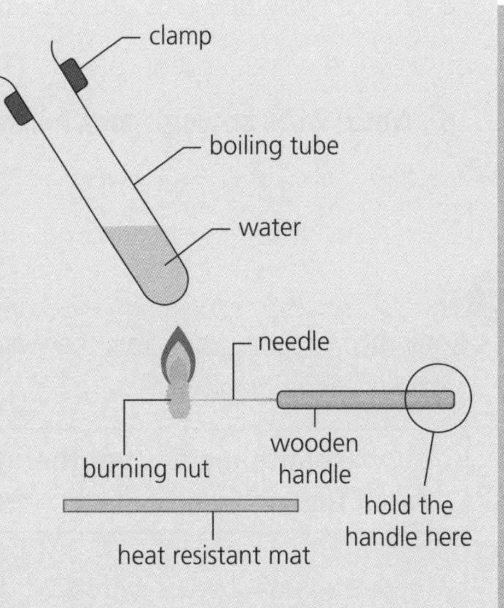

The student works out from the temperature rise that there are 25 000 J in 1 g of peanuts. He realises that not all the energy from the food is heating up the water.

c Why does not all of the energy from the food heat up the water?

d Is the number that he has worked out bigger or smaller than the actual value of the energy stored? Explain your answer.

1 Match the beginning of each sentence with the correct end of the sentence. Connect the boxes together with lines.

The process that converts light energy from the Sun into chemical energy is …
Wind turbines turn …
Energy in our food …
When it rains, water in rivers flows into artificial lakes that …
Energy in fuels comes from the Sun …
Solar panels use energy directly from the Sun …

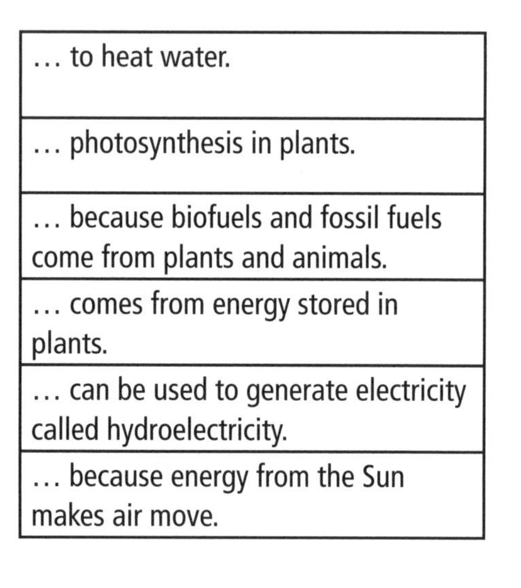

… to heat water.
… photosynthesis in plants.
… because biofuels and fossil fuels come from plants and animals.
… comes from energy stored in plants.
… can be used to generate electricity called hydroelectricity.
… because energy from the Sun makes air move.

2 A student is tracing the energy in fuels and food back to the Sun. Complete the diagrams by filling in the boxes.

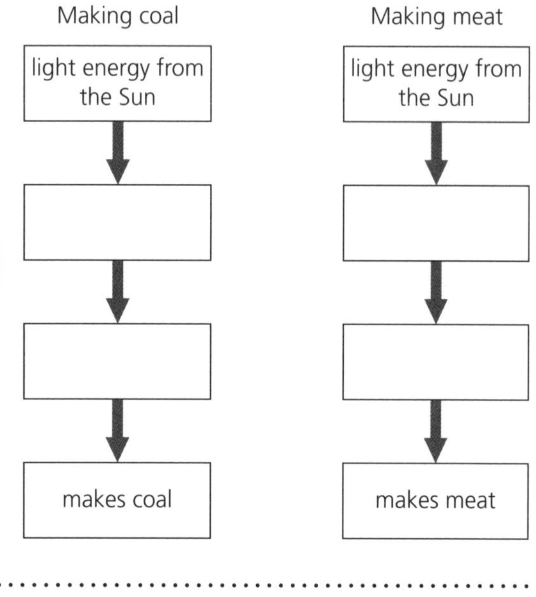

3 We can use energy from the Sun directly or indirectly to generate electricity.

solar cells biofuels gas petrol solar panels

a Which word or words are a *direct* method of generating electricity?

………………………………………………

b Which word or words are an *indirect* method of generating electricity?

………………………………………………

c Which word or words are *not* a method of generating electricity?

………………………………………………

d Which word or words took millions of years to form?

………………………………………………

E

Copy the Venn diagram. Sort the words in the box into the correct sections of the diagram.

solar panels geothermal hydroelectricity fossil fuels wind power food solar cells biofuels

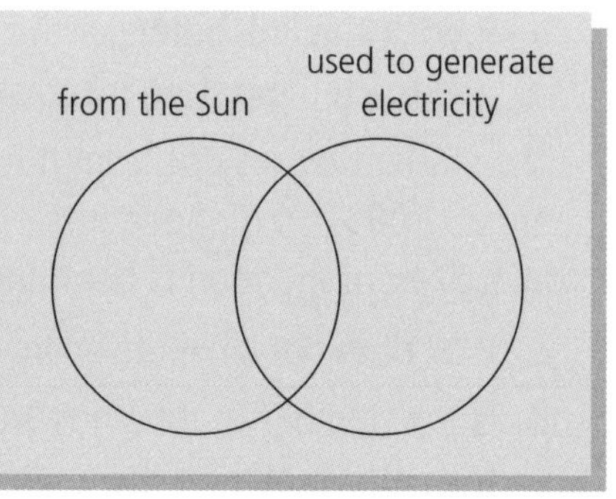

2.3 Energy types

1 a Complete the table about energy types below using the words from the list.

moving | vibrating | electrical | light | chemical | springs | thermal | lifted up

Type of energy	Definition / source
Gravitational potential energy	Objects .... have this.
Kinetic energy	.... objects have this.
Sound energy	.... objects produce this.
.... energy	The Sun produces this.
.... energy	Hot objects have a lot of this.
Elastic potential energy	.... store this.
.... potential energy	Found in food and fuels.
.... energy	How energy is transferred in circuits.

b Which types of energy are stores?

....

c Which types of energy are ways of transferring energy?

....

2 Read the information in the box then answer the questions below.

> Mina gets on her bike to meet her friend at the cinema. They walk upstairs to find their seat. The music at the start of the film is very loud. They enjoy seeing the people in the film. They go downstairs and leave the cinema. On the way home they buy some food.

Each sentence talks about a type or types of energy. What are they?

a ....

b ....

c ....

d ....

e ....

f ....

E

a A student is making notes about nuclear energy but is confused. Write out a correct version of his notes.

1 The reaction that produces energy in the Sun is a chemical reaction.

2 Nuclear fusion happens when uranium breaks down.

3 Nuclear fission happens when uranium combines.

4 The fuel in a nuclear power station is hydrogen gas.

b Explain why the Sun cannot be a huge ball of fire.

2.4 Energy transfer

1 Which of these statements about energy transfer diagrams are correct? Write ***T*** next to the statements that are true. Write ***F*** next to the statements that are false.

- **a** They show how energy is transferred in an object or process.
- **b** They show how energy is lost or wasted.
- **c** They show the different types of energy involved in a process.
- **d** They show how much energy is used.

2 Fill in the boxes to show the energy transfers.

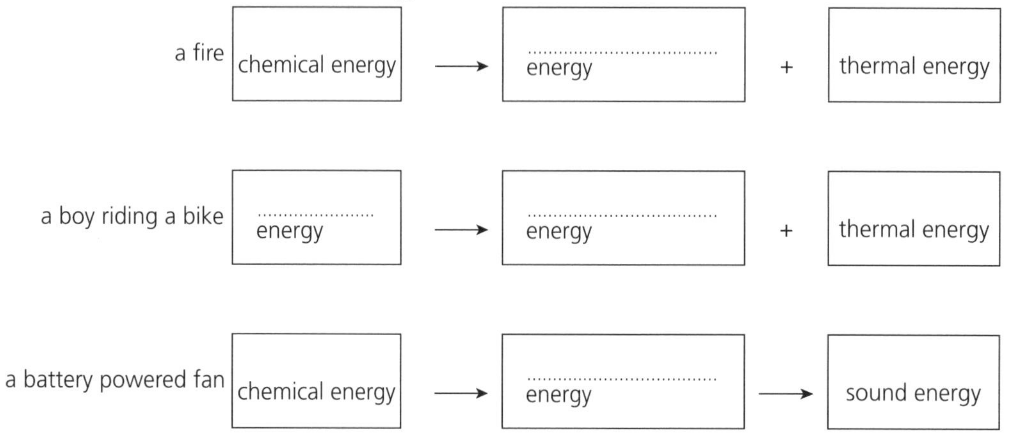

3 Choose from this list of objects to complete the energy transfer diagrams below. You do not need to use all the words.

kettle   television   plant   radio (battery powered)   car

.... light energy → chemical energy

.... chemical energy → sound energy + thermal energy

.... electrical energy → sound energy + thermal energy

E

- **a** Draw energy transfer diagrams for the following processes.
  - You eat your breakfast and walk to school.
  - You walk up a hill and back down again.
  - A candle burning.
  - A loudspeaker in a television.
- **b** In which of the processes above does it make more sense to talk about energy 'transfer'?
- **c** In which of the processes above does it make more sense to talk about energy 'changes'?
- **d** Name a process in which thermal energy is *not* produced.

1 Use the words and phrases from the box to complete the sentences below. Use each word once, more than once, or not at all.

transferred	money	conservation	protection	the same
a different	destroyed		types	created

Energy cannot be .... or .... , it can only be ....

This is the law of .... of energy. Energy is a bit like .... If you know how much you have to begin with you know that you will have .... amount at the end. The different .... of energy are ways of keeping track of where the energy is.

2 Circle the correct word or phrase in each **bold** pair in the sentences below.

a In a loudspeaker the **useful / wasted** energy is sound.

b In a fan the **useful / wasted** energy is sound.

c In a motor the **useful / wasted** energy is thermal.

d In an oven the **useful / wasted** energy is thermal.

3 Here are some diagrams of electrical appliances.

a Fill in the percentage of energy that is wasted.

80% of the energy is used to heat the water

................% is wasted.

50% of the energy is given out as sound

................% is wasted.

b Suggest one way that energy is wasted in each device.

....

....

c Which device is the most efficient?

....

E

a Use the law of conservation of energy to fill in the table below for four machines: A, B, C, and D.

Machine	Useful energy (J)	Wasted energy (J)	Total energy (J)
A	25	5	
B	2000		2500
C		1500	2500
D	20		30

b Explain how you used the law of conservation of energy to fill in the table.

c Is machine A or D *more* efficient? Explain your answer.

d Is machine B or C *less* efficient? Explain your answer.

1 Complete the table by placing a tick (✓) in the correct column or columns for each statement.

	Gravitational potential energy	Kinetic energy
Energy that something has because of its position.		
Energy that something has because of its movement.		
This gets bigger if an object is higher off the ground.		
Is measured in joules.		
A fast-moving elephant has lots of this.		
A walking mouse has less of this than the elephant has.		

2 A student makes a track for marbles to roll down. He lets the marble go from point A.

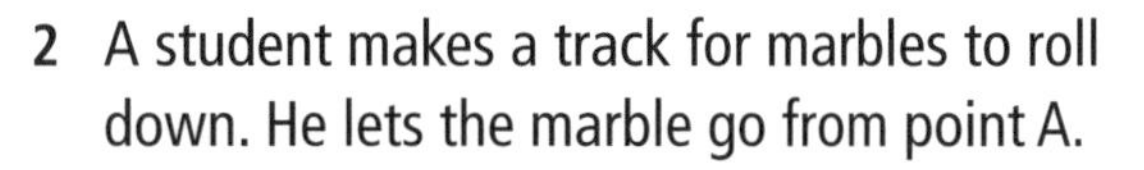

a At which point A, B, C, D, E, or F does the marble have the *most* gravitational potential energy?

....

b At which point A, B, C, D, E, or F does the marble have the *no* gravitational potential energy?

....

c At which point A, B, C, D, E, or F does the marble have the *most* kinetic energy?

....

d Could the top of the loop (C) be higher than point A? Explain your answer.

....

e At point F the marble does not have the same amount of energy as it had at the start. Where has the energy gone?

....

3 A student has dropped a ball from different heights into wet sand.

a Describe the energy changes from the point when he drops the ball to the point where the ball is in the sand and has stopped.

....

....

Here are some values for the gravitational potential energy and kinetic energy at different points between his hand and the floor. They are not in the correct order.

Point	GPE (J)	KE (J)
A	0	1
B	0.75	0.25
C	1	0
D	0.25	0.75

b Which point A, B, C, or D is the floor? ....

c Which point A, B, C, or D is his hand? ....

d At which point A, B, C, or D is the ball travelling the fastest? ....

1 Write ***T*** next to the statements that are true. Write ***F*** next to the statements that are false. Then write the corrected versions of the statements that are false.

a When materials deform they store chemical energy.

b If a spring is stretched more, it will store more elastic potential energy.

c The springs in a mattress will store less elastic potential energy if someone heavier sits on it.

d If something does not return to its original shape when we remove the force, we say it is elastic.

e When materials deform, the energy wasted is thermal energy.

f The muscles in your legs act like elastic bands when you walk.

Corrected versions of false statements:

....

....

....

2 A student is collecting data about the energy stored in different elastic bands used in a catapult.

A student pulls back each band by the same amount. She lets go of the ball. She measures how far the ball travels.

Elastic band	Distance travelled (m)
A	1.50
B	2.50
C	1.25
D	1.20

a What are the energy changes from the point when she lets go of the elastic band?

....

b Put the bands in order of stored energy starting with the band that stores the least.

....

c If the band was pulled back more, what would happen to the distance the ball travels?

....

E

A student completes an investigation into bouncing balls. He changes the height that he drops the ball from and measures the height to which it bounces. Here are his results.

Height of drop (cm)	Height of bounce (cm)
50	40
100	70
150	90
200	105
250	120

a Explain why the ball does not bounce back to the same height.

b After the first drop the student expected the next bounce to be 80 cm. Suggest a reason why the student expected it to be 80 cm. When you double the height of the drop does the height of the bounce double, more than double or less than double?

2.8 Suggesting ideas – energy in fuels

1 Here are the stages for doing a practical investigation. They are in the wrong order. Write the letters in the correct order.

A Collect data.

B Check the data to see if they agree with the explanation.

C Ask a question.

D Suggest an explanation using scientific knowledge.

E Present the data in a table, graph, or chart.

2 Here are some data from an investigation into biofuels.

Biofuel	Volume of biofuel used (cm^3)	Volume of water heated (cm^3)	Temperature rise (°C)
dry wood	1	10	35
ethanol	1	10	65
green wood	1	10	20
methanol	1	10	80

a How does the table show that the student did a fair test?

....

b Put the fuels in order of most energy stored to least energy stored.

....

3 A student wants to investigate the different types of ball in sport and how they bounce.

a Change the boy's idea into a question that he can investigate that starts 'How does…'

....

b Suggest an explanation for why different balls might bounce differently.

....

c On a piece of paper or in your books draw a table that the student could use to collect data.

E

Here are the headings of some tables that students are going to use to collect data.

Investigation A

Volume of liquid fuel used (cm^3)	Temperature change of the water (°C)

Investigation B

Type of surface	Height of bounce of a ball (cm)

For investigation A and investigation B:

a Suggest the idea that the student is going to test in each case.

b Suggest what things need to be kept the same in each case.

2.9 Suggesting ideas – field studies and models

1 Draw lines to match the methods for answering scientific questions to the definitions.

Methods
field study
regular observations
make a model
practical investigation

Definitions
Make observations over a period of time, or in lots of places at the same time.
Make observations of organisms in their natural habitat.
Collect data or make observations in a laboratory.
Use a computer model or physical model.

2 Here is a list of questions. Suggest which method could be used to try to answer it.

a How has the temperature of the sea changed? ....

b Do lions kill more animals in the winter than the summer? ....

c How will the weather change in the future? ....

3 Here are some data that scientist have collected about the levels of carbon dioxide in the atmosphere. They were collected in Hawaii every November for over 60 years. Carbon dioxide levels are measured in parts per million or ppm.

Year	CO_2 level (ppm)
1960	315
1970	324
1980	337
1990	353
2000	368
2010	388

a Suggest which method scientists used to collect this data.

....

b Scientists want to use these data to predict what might happen in the future. Which method could they use to do that?

....

c What do the data show is happening to the levels of carbon dioxide in the atmosphere?

....

d Do these data suggest that we should not burn fossil fuels any longer? Explain your answer.

....

....

E

Some scientists drill down into the ice in Antarctica to find ice containing air trapped from millions of years ago. They analyse the air in a laboratory. Here are some of the data that they collected.

Year before now	CO_2 level (ppm)
100 000	225
80 000	220
60 000	190
40 000	180
20 000	200
0	388

a Why is it difficult to choose which of the methods they have used to collect the data?

b Is there a pattern in the data? Explain your answer.

c What is the question that scientists can answer with these data?

d Suggest a question that scientists *cannot* answer with these data.

2.10 Energy calculations and Sankey diagrams

1 Complete the equations for calculating efficiency by filling in the gaps.

efficiency (..................) = $\dfrac{\text{.................. energy out (..................)}}{\text{.................. energy supplied (..................)}}$ × 100 ....

2 Here are some data about some devices that transfer energy.

Use the law of conservation of energy to complete the table, then calculate the efficiency of each device. Round up your answer to the nearest whole number.

Device	Useful energy (J)	Wasted energy (J)	Total energy (J)	Efficiency
light bulb	5		30	
kettle		500	2000	
television	2500	2500		
car	100		400	

3 A student makes some notes about the energy changes in a new energy-saving lamp. He calculates the efficiency of the lamp:

Energy in = 10 J　　　Useful energy = 2 J　　　Efficiency = $\frac{10}{2}$ × 100 = 500

a How can you tell from the efficiency that the student has *not* done the calculation of efficiency correctly?

....

b What other mistake has the student made?

....

c Write out the calculation correctly in the space below.

4 A student draws a Sankey diagram for her music player. Each square represents 1 J.

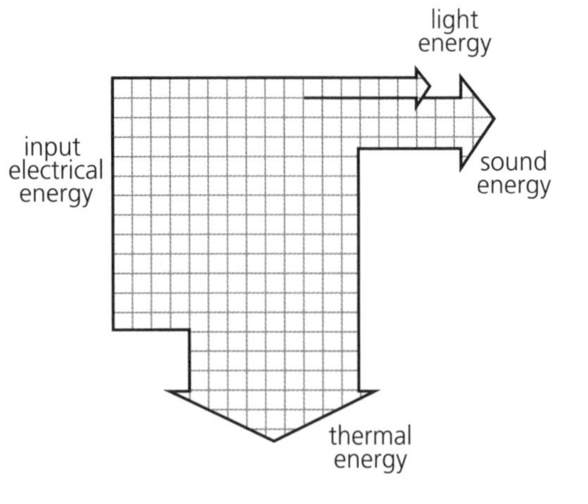

a Use the information on the diagram to calculate the total energy in.

....

b Use the information on the diagram to calculate the useful energy out.

....

c Calculate the efficiency of the music player. Round the number to the nearest percent.

....

....

d What percentage of the energy in the music player is wasted?

....

3.1 The night sky

1 There are many objects that you can see in the night sky. Write the letter for each object in the list below in the correct part of the Venn diagram.

A planet
B moon
C comet
D asteroid
E space station

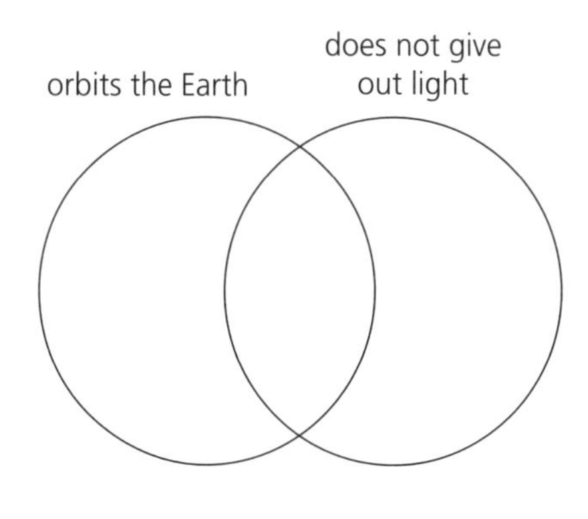

2 Write ***T*** next to the statements that are true. Write ***F*** next to the statements that are false.

Then write the corrected versions of the statements that are false.

a All the objects that we see in the night sky are in orbit around the Sun.
b The stars in the night sky are smaller than our Sun.
c Planets appear to wander across the night sky.
d The planets we can see without a telescope are Mercury, Venus, Mars, and Jupiter.
e We cannot see any man-made objects in the night sky.

Corrected versions of false statements:

....

....

3 Here are some facts about asteroids, comets, and meteors. Tick (✓) the correct column or columns.

	Comet	Meteor	Asteroid
made of ice and dust			
made of rock			
called 'shooting stars'			
visible in the night sky			
burns up as it enters the atmosphere			

The time it takes for a comet to return is called its period. Here are some data about two comets.

	Comet Hale-Bopp	Halley's Comet
period	2500 years	76 years
last seen	1997	1986
furthest distance from the Sun (compared with distance from Earth to Sun)	370	35
closest distance to the Sun (compared with distance from Earth to Sun)	0.9	0.6

a When will the comets next be seen?
b What is the link between the period of the comet and the furthest distance from the Sun?
c Which comet comes closest to the Earth? Explain your answer.

3.2 Day and night

1 Look at the diagram opposite.

a Label the side of the Earth where it is daytime with the word 'day'.

b Label the side of the Earth where it is night-time with the word 'night'.

c Label the Earth's axis with the word 'axis'.

d Label the north pole with the word 'north'.

e Label the south pole with the word 'south'.

2 Here is a diagram showing the Sun's path across the sky.

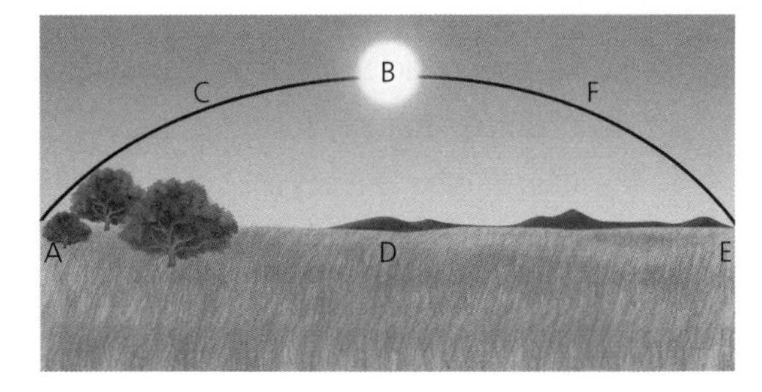

For each of the questions below write the correct letter.

a Where does the Sun rise? ………………

b Where does the Sun set? ………………

c Where is the Sun at midday (noon)? ………………

d Which direction is South? ………………

e Where is the Sun in the morning? ………………

f Where is the Sun in the afternoon? ………………

3 Use the words and phrases from the box to complete the sentences below. Use each word once, more than once, or not at all.

orbit	spins	day	anticlockwise	move	24 hours	night	year	365 years	clockwise	spin

a The Sun appears to …………………………… across the sky each …………………………… .

b The Earth …………………………… on its axis in …………………………… . We call this time one …………………………… .

c If you are on the side of the Earth that is facing the Sun it is …………………………… and if you are on the side of the Earth facing away from the Sun it is …………………………… .

d The Earth spins …………………………… as viewed from above.

E

Jengo and Simba are modelling the Earth and the Sun for the other students in their class. They want to explain why we get day and night. Jengo wraps a big map of the world around Simba and moves the map so that Africa is at the front. Simba holds the map in place.

Simba stands with his back to Jengo. Jengo takes a big torch and shines it at Simba.

a Explain how Simba and Jengo can explain day and night with this model.

b A student thinks that the Earth is not spinning because she cannot feel it moving. How would you convince her that it is?

3.3 The seasons

1 Look at the diagram below.

a What do we call the circular path that the Earth takes as it moves around the Sun?

....

b How do you know that the diagram is not to scale?

....

c Fill in the table below to show the seasons in the southern and northern hemispheres.

Position	Southern hemisphere	Northern hemisphere
A		
B		
C		
D		

2 Circle the correct word or phrase in each **bold** pair in the sentences below.

a When the northern hemisphere is tilted towards the Sun it is **summer / winter**. The **days / nights** are longer than the **days / nights** and the Sun is **high / low** in the sky at noon.

b When the southern hemisphere is tilted away from the Sun it is **summer / winter**. The **days / nights** are longer than the **days / nights** and the Sun is **high / low** in the sky at noon.

c The weather is **warmer / colder** in the summer because the Sun is shining for a **longer / shorter** time and has **more / less** time to warm up the air and ground. The Sun is higher in the sky so its rays **are / are not** spread out.

3 Here is some information about sunrise times in a city.

	Time of sunrise
1 January	05:30
1 April	07:00
1 July	08:30
1 October	07:00

a Is the city in the northern or the southern hemisphere?

....

b Explain how you worked out the answer to part **a**.

....

....

E

A teacher is showing her class why it is hotter in the summer than the winter. She gets two trays of sand and two lamps and sets them up like this.

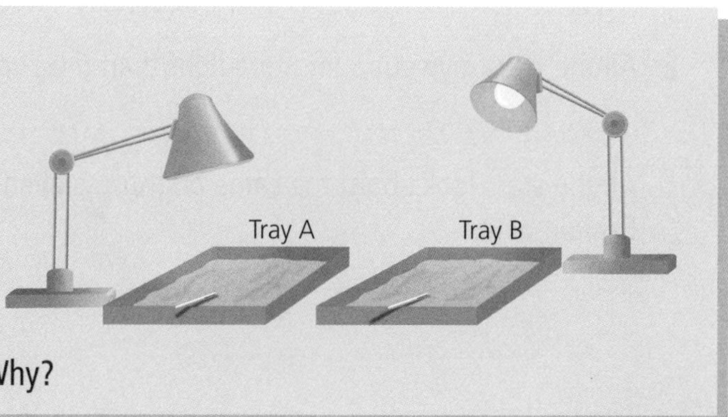

a In which tray will the thermometer show the highest temperature?

b How is this demonstration a good model for why it is hotter in summer than winter? Why?

1 Write ***T*** next to the statements that are true. Write ***F*** next to the statements that are false. Then write the corrected versions of the statements that are false.

a You see the same stars all the year round.

b People in the southern hemisphere and northern hemisphere see different stars.

c Our star is the brightest star.

d The stars in a constellation are all about the same distance away.

Corrected versions of false statements:

....

....

2 Here is a photograph that shows the paths of the stars in the sky at night.

a Explain why the paths of the stars are circles.

....

b Name the star nearest the centre in the southern hemisphere.

....

c Name the star at the centre in the northern hemisphere.

....

d Why did sailors use these stars to navigate?

....

3 This is the constellation Cassiopeia.
Here is some information about the stars in Cassiopeia.

	Light they give out compared to the Sun (Sun = 1)	Distance to the star
A	1000	442
B	60	99
C	4200	613
D	600	228
E	30	54

a Cassiopeia is a constellation. What is a constellation?

....

b All the stars give out a lot more light than the Sun. Why do they not look as bright as the Sun?

....

c All the stars look about the same brightness even though they give out different amounts of light. Explain why.

....

....

3.5 Our Solar System

1 Tick the boxes for the correct statements about planets.

	Correct?
All planets orbit the Sun.	
Pluto is an exoplanet.	
Dwarf planets do not orbit the Sun.	
Planets that are further from the Sun are colder than planets that are closer.	
All the inner planets are rocky.	
Asteroids orbit the Sun between Mars and Jupiter.	

2 Here is some information about the planets of the Solar System. All of the data show numbers compared with the Earth.

	Mercury	Venus	Earth	Mars	Jupiter	Saturn	Uranus	Neptune
distance from Sun	0.3871	0.7233	1	1.524	5.203	9.539	19.19	30.06
time to orbit Sun	0.24	0.62	1	1.88	11.86	29.46	84.01	164.79
diameter	0.4	0.9	1	0.5	11.2	9.4	4.0	4.0
mass	0.06	0.82	1	0.11	317.89	95.18	14.53	17.14
number of moons	0	0	1	2	>28	30	24	8

Here are the planets put into different order. Identify the order.

a Mercury, Mars, Venus, Earth, Neptune, Uranus, Saturn, Jupiter = order of ....

b Mercury, Venus, Earth, Mars, Neptune, Uranus, Jupiter, Saturn = order of ....

c Which two quantities would produce the same order? Why?

....

....

E

There are different types of planet. Explain the difference between a planet and an exoplanet.

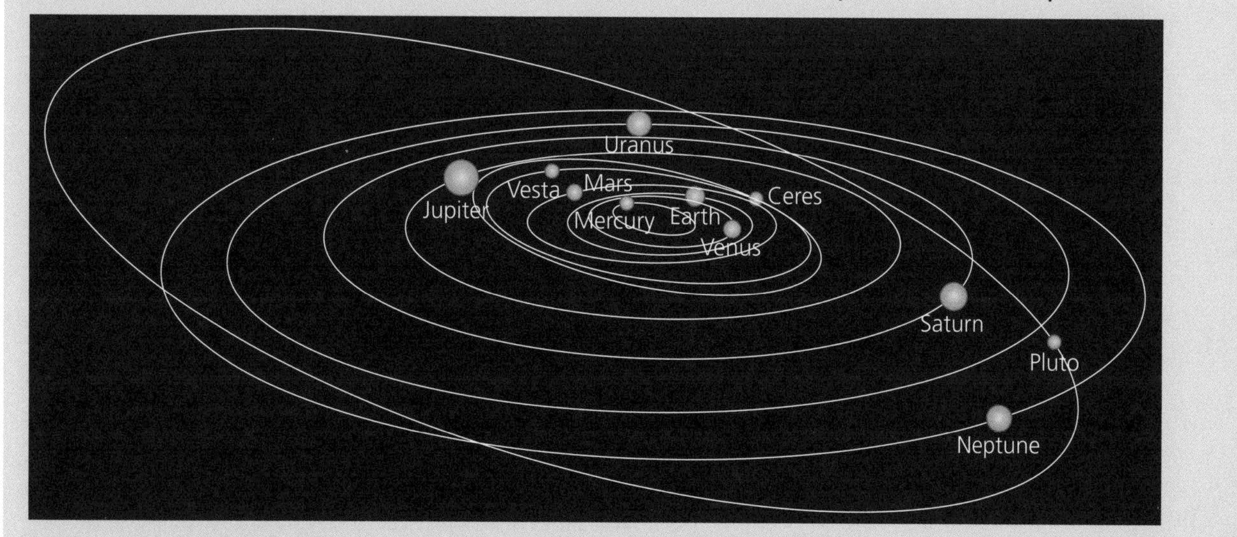

Use the diagram to explain:

i why someone might think that Pluto *was* a planet

ii why someone might think that Pluto *was not* a planet.

3.6 The Moon

1 Here are some pictures of the phases of the Moon.

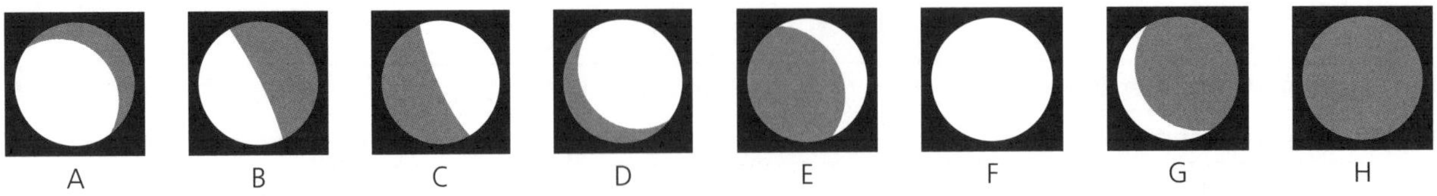

A B C D E F G H

a Put the phases in order by writing the letters in the correct order, starting with the full moon.

....

b Which letter or letters show a crescent moon?

....

c Which letter shows a new moon?

....

d A student says that only the part of the Moon's surface that we see is actually lit up by the Sun. How would you correct them?

....

....

2 This is a diagram showing an eclipse.

a Label the diagram.

b Is this a diagram of a lunar eclipse or a solar eclipse? Explain your answer.

....

....

....

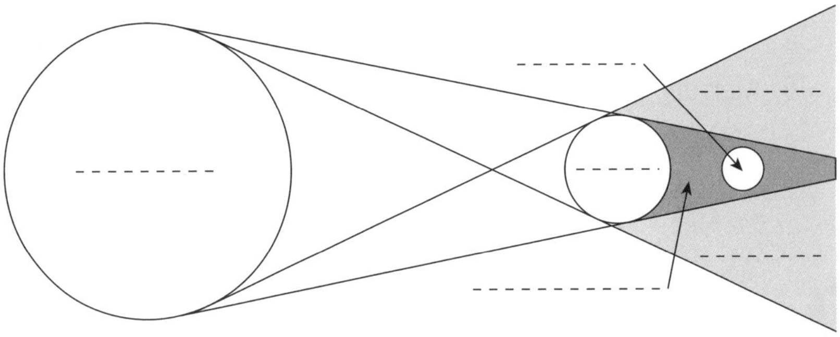

c In the space below draw and label a diagram to show the other type of eclipse.

3.7 Explanations – geocentric model

1 Use the words and phrases from the box to complete the sentences below.
Use each word once, more than once, or not at all.

explanations	measurements	evidence	model	questions	stories

From the first time that people have looked up at the night sky they have asked .... about what they saw. They used .... to explain what they saw.

Scientists make .... of things that they see. These .... and the .... that they make provide evidence. They use the evidence to develop .... to account for the evidence. Sometimes they make a .... and use it to explain what they see.

2 Here is a diagram of the geocentric model but the planets, Moon, and Sun are missing.

a Add the planets, Moon, and Sun to the diagram and label them.

b Where were the stars in this model?

....

c What was the main piece of evidence for the geocentric model?

....

3 a Explain the difference between the stories that people used to make up about the objects in the night sky and models such as the geocentric model.

....

b Why was the motion of the planets a problem for the geocentric model?

....

c How did Ptolemy adapt the geocentric model so he could account for the motion of the planets?

....

d Why was Ptolemy's explanation so useful, and used for so many years?

....

E

a Explain why Greek astronomers could not detect the motion of the stars during the year.
Draw a diagram and use it to explain your answer.

b Why can astronomers detect that motion now?

3.8 Explanations – heliocentric model

1 Galileo made observations of the phases of Venus.
These observations were made by in 1610 using a telescope.

a Apart from seeing the phases how does the appearance of Venus change in this diagram?

....

....

At the time astronomers were using a geocentric model of the Solar System.

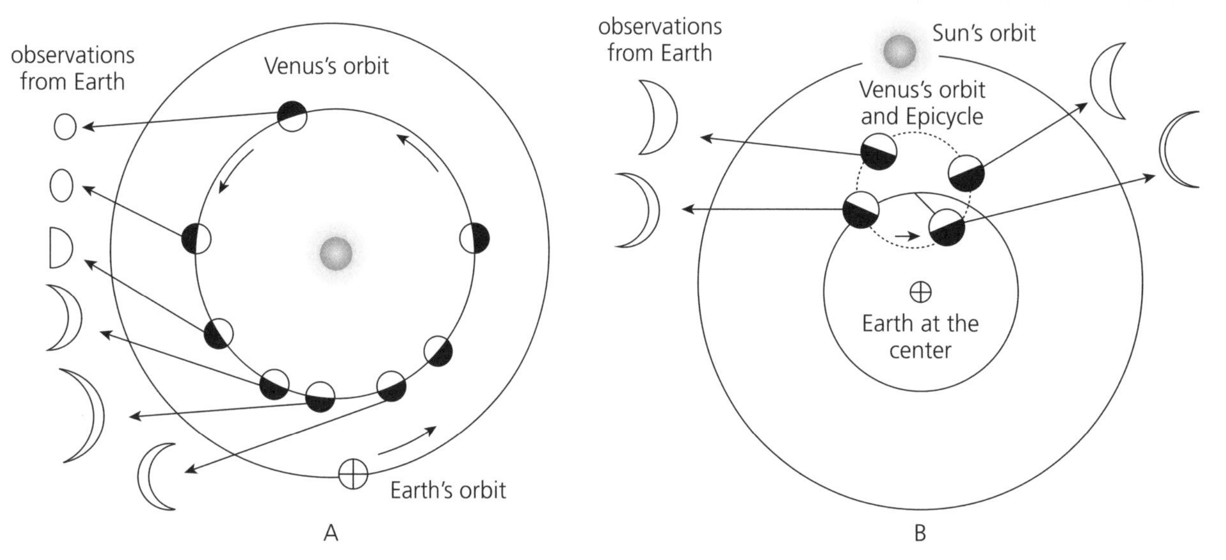

b Which of the diagrams above shows the geocentric model?

....

c Which model best explains the changing appearance of Venus in your answer to part **a**? Explain your answer.

....

....

2 This table compares Copernicus and Galileo. Put a tick (✓) in the correct column or columns for each statement.

	True for Copernicus	True for Galileo
wanted to talk his about ideas		
thought that the Sun was at the centre of the Solar System		
used a telescope to make observations		

E

Here is a list of observations that provide evidence for models of the Solar System. Sort the observations into those that are explained *better* by the geocentric model and those that are explained *better* by the heliocentric model.

a phases of Venus
b objects fall toward the centre of the Earth
c the stars don't seem to move during the course of a year
d Jupiter has moons
e the movement of the planets across the sky

3.9 Communicating ideas

1 Look at the information in the table about Aryabhata's explanations of some observations. Answer the questions below.

What we see	Explanation at the time	Aryabhata's explanation
movement of the Sun and stars across the sky	The sky rotates.	The Earth spins on its axis.
motion of the planets	The planets move around the Earth.	The planets move on epicycles on orbits around the Earth.
solar and lunar eclipses	Gods swallow the Sun during a solar eclipse and the Moon during a lunar eclipse.	Eclipses are caused by the shadow of the Moon or the Earth.

a Which observation or observations does Aryabhata correctly explain?

....

b Which observation or observations does Aryabhata not correctly explain?

....

c Aryabhata lived over a thousand years ago. Suggest two reasons why his correct explanations weren't widely known.

....

....

2 The astronomers in ancient Egypt used lots of astronomical instruments.

a What did they use a merkhet for?

....

b How did they measure time?

....

c What is the difference between the observations that they made and the drawings that ancient people made?

....

....

E

Here is a timeline showing how the heliocentric model of the Solar System developed.

Years ago	Where	Idea
2860	India	The Earth is in motion around the Sun.
2260	Greece	Aristarchus describes the heliocentric model in a book.
1862	Greece	Ptolemy publishes his geocentric model.
1000	Iraq	Al-Haytham writes about Ptolemy's model.
710	Baghdad	Al-Shatir improves Ptolemy's model.
469	Poland	Copernicus publishes the heliocentric model.

a Suggest a reason why it took over 800 years for Islamic astronomers to write about Ptolemy's model.

b Suggest why the Islamic astronomers did not work on the heliocentric model.

3.10 Beyond our Solar System

1 Imagine that you could get in a spaceship and travel away from the Earth. Here is a list of things that you would pass.

- **A** the edge of the Solar System
- **B** the Andromeda galaxy
- **C** the Oort cloud
- **D** the Kuiper belt
- **E** the asteroid belt
- **F** the edge of the Milky Way
- **G** Proxima Centauri

a Write the letters in the order that you would see them.

b Which object is *inside* the Solar System?

....

c Why is the edge of the Universe not on the list?

....

2 A student draws a diagram to show where things in the Universe are. She wants to place each object on the list into one of three circles.

- **A** Earth
- **B** galaxy – not the Milky Way
- **C** comet
- **D** Oort cloud
- **E** Proxima Centauri
- **F** star – not the Sun
- **G** Moon
- **H** Kuiper belt

inside the Universe but outside the Milky Way

inside the Milky Way but outside the Solar System

inside the Solar System

a Why has the student put the three circles inside each other?

....

b Put the letters of each of the objects in the correct place on the diagram. One object is difficult to put into a circle. Which one, and why?

....

c Why would it not be possible to add 'planet' to the list?

....

3 A photographer has taken a photograph of a very large crowd of people. He wants to work out how many people are in the picture.

a How could he do that (approximately) without counting each one individually?

....

....

b How is the method that he uses similar to the way that astronomers count stars?

....

....

3.11 Using secondary sources

1 This is a list of places where you can find data about the temperature at noon every day. Some sources are primary sources and some are secondary sources. Tick (✓) the correct column.

	Primary source	**Secondary source**
data from book in a library		
data from measurements that you have made		
data from a field study		
data on a website		
data from measurements that someone else has done and given to you		

2 Here is some data about galaxies that you can see without a telescope. The information comes from a website.

Galaxy	**Distance (light years)**	**Type**	**Brightness (Sun = 1)**
large Magellanic cloud	160 000	irregular (no shape)	50
small Magellanic cloud	200 000	irregular (no shape)	10
Andromeda	2 500 000	spiral	5
Omega Centauri	18 000	spherical	4
Triangulum	2 900 000	spiral	0.6
Centaurus A	13 700 000	elliptical	0.08
Bode's galaxy	12 000 000	spiral	0.04

a Are the data primary or secondary data?

....

b A student thinks that spiral galaxies are brighter than elliptical galaxies. Is she correct? Explain your answer.

....

....

c Another student thinks that galaxies that are closer are brighter.

i Plot the distance and brightness of the first four galaxies in the table on the graph paper. Plot an *x*-axis from 0 – 3 000 000 light years, and a *y*-axis from 0–60 for brightness.

ii Does the graph show a link between brightness and distance?

....

iii What could you do to be surer of your answer to part ii?

....

E

a All scientists check their data by repeating their experiments. Why?

b If you are looking at data that you have found in a secondary source it is important to check them. Why?

3.12 The origin of the Universe

1 Write ***T*** next to the statements that are true. Write ***F*** next to the statements that are false. Then write corrected versions of the statements that are false.

a The Big Bang theory says that the Universe expanded 14 billion years ago from something smaller than an atom.

b The Earth is the centre of the Universe.

c When astronomers look at galaxies they are all moving towards us.

d The Solar System formed about 10 billion years ago.

e An alternative to the Big Bang theory is the Solid State theory.

Corrected versions of the false statements:

....

....

2 A student is making a timeline about the history of the Universe. He takes a strip of paper that is 1 m.

What's happening	Time (million years ago)
the Big Bang	14 000
Solar System formed	5000
Earth cooled	4200
single-celled organisms	3900
oldest rock	3100
multicellular organisms	630
land plants	550
dinosaurs lived	200
dinosaurs died out	65
humans exist	0.5
today	0

a Which events will the student find it difficult to write on his timeline? Why?

....

....

....

b Another student says that he can solve the problem by having piece of paper that is twice as long. Is he correct? Explain your answer.

....

....

3 A student wants to make a model of the expanding Universe. He takes an elastic band and cuts it so that he has a long piece of elastic.

a Explain how he can use the elastic to model the expanding Universe.

....

....

b How could he use the elastic band to show that galaxies that are further away are moving faster?

....

....

c What is the evidence that the Universe is expanding?

....

....

1 Write ***T*** next to the statements that are true. Write ***F*** next to the statements that are false. Then write the corrected versions of the statements that are false.

a To calculate speed you need to know distance and force.

b If you use the total distance and total time, then the speed you can calculate is the average speed.

c Speed is measured in units such as kilometres per hour (km/h) and newtons per second (N/s).

d A steady speed is a speed that is changing all the time.

Corrected versions of false statements:

....

2 A cyclist takes 16 seconds to travel 200 metres.

a Calculate her speed. Is this the speed at the start, at the finish, or the average?

....

b Explain your answer.

....

3 Read the report below.

Some of the world's fastest athletes competed at the Summer Olympic Games in London in August 2012. One of these athletes, David Rudisha of Kenya, ran 800 metres in 1 minute 41 seconds to win the gold medal. This was a new world record. In the same Olympic Games teenager Kirani James of Grenada won a gold medal by running 400 metres in 44 seconds.

a How long did David Rudisha take to run 800 m in seconds?

....

b What was David Rudisha's average speed for the 800 m? (Round the number up to 2 decimal places.)

....

c What was Kirani James's average speed for the 400 m? (Round the number up to 2 decimal places.)

....

d If Kirani ran the 800 m at the same average speed as the 400 m, how long would it take him?

....

E

Here are some data about the land speed record. This is the highest speed achieved by a wheeled vehicle on land. The speed of the cars is measured over 1600 m.

a How long would it take a car from 1898 to travel 1600 m?

b How long would it take a car from 1928 to travel 1600 m?

c When did the cars start to use rockets rather than engines? How can you tell from the data?

Year	Speed (km/h)	Speed (m/s)
1898	63	18
1906	205	57
1928	334	93
1947	394	109
1964	403	112
1965	893	248
1983	1020	283
1997	1228	341

4.2 Taking accurate measurements

1 Two students are dropping paper cups in an experiment about weight and air resistance. One student uses a stop-clock. The other student uses timing gates. Here are their results.

Results for student 1:

Number of cups	Time to fall (s)
1	0.73
2	0.62
3	0.54
4	0.43

Results for student 2:

Number of cups	Time to fall (s)
1	0.7
2	0.6
3	0.5
4	0.4

a Which student used the timing gates? How can you tell?

....

b Which student will have results that are more accurate? Why?

....

c Here are some times. Put them in order from the most precise to the least precise.

1.45 1.4 1.452 1

....

2 A teacher is timing a race between two students. She starts her stopwatch when she hears the gun. She stops her stopwatch when she sees the winner cross the line.

a Why will it be difficult for her to get an accurate reading of the winner's time?

....

....

b Why will it be difficult for her to get a precise reading of the winner's time?

....

....

3 Time-lapse photography means taking pictures at certain time intervals. This is how speed cameras work. There are lines painted on the road and the speed camera takes two pictures very close together. Using the lines on the road you can work out distance.

a What else do you need to know to calculate the speed?

....

....

b If a speed camera was not working properly and took photographs too far apart, what would be the problem with working out the speed?

....

....

c There are two speed cameras, one in an area where the speed limit is 30 km/h and another in an area where the speed limit is 50 km/h. In which one will there be a smaller time interval between the photographs? Explain your answer.

....

....

1 Look at the graphs A to C below.

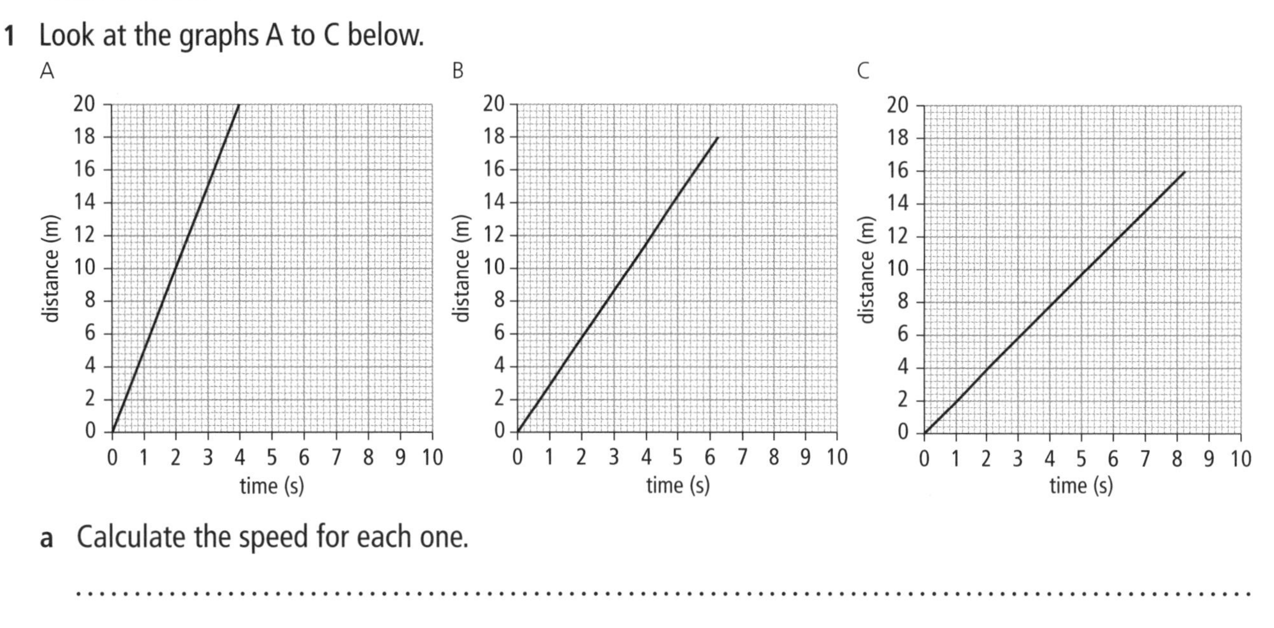

a Calculate the speed for each one.

....

....

....

....

b How can you tell which speed is fastest without doing a calculation?

....

2 A boy races his bicycle along a track. Use this graph to answer the questions.

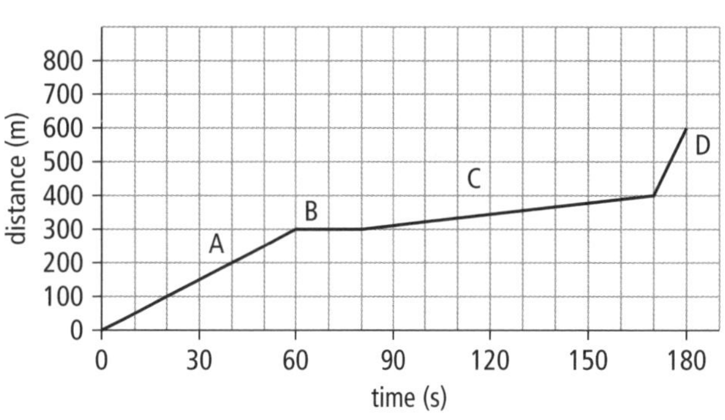

a How many metres did he travel in total?

....

b What was the total time for the journey in seconds?

....

c What was the average speed for the journey in metres per second?

....

E

A student drops a stone off a tall cliff.

Look at the graph of distance against time for the stone.

a How far had the stone fallen after 2 seconds?

b How can you tell from the graph that the stone is falling faster and faster?

c How would the graph be different if the student dropped a stone with a bigger area? Draw a line on the graph to show what it would look like.

d Explain why you have drawn the graph that you have drawn.

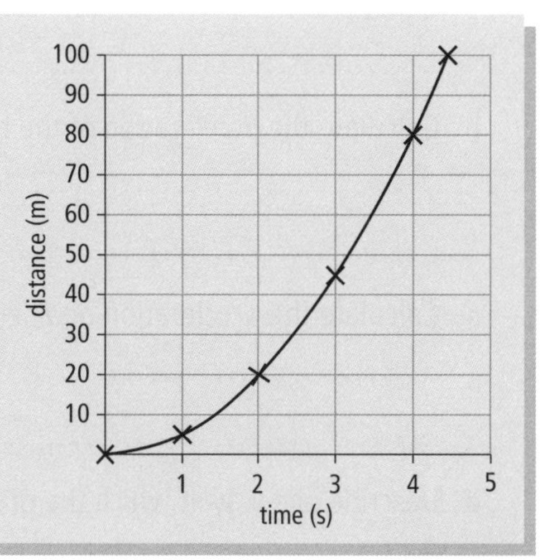

4.4 Acceleration and speed–time graphs

1 A car's speed changes from 6 m/s to 18 m/s in 2 seconds. Write ***T*** next to the statements that are true. Write ***F*** next to the statements that are false.

a The acceleration is 12 m/s^2.

b The deceleration is 6 m/s^2.

c The acceleration is –6 m/s^2.

d The acceleration is 6 m/s^2.

2 Look at the speed–time graphs below.

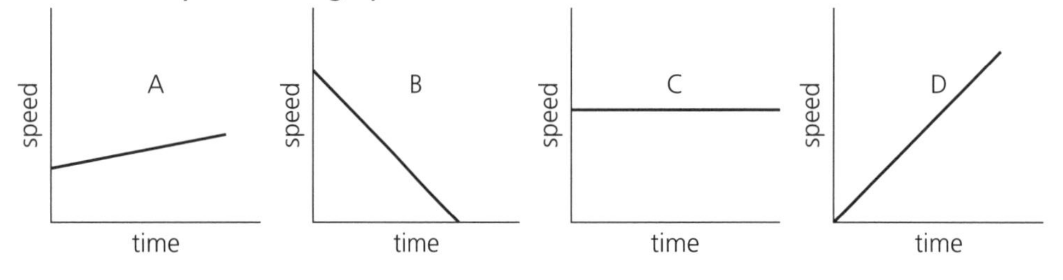

In which diagram or diagrams is the car:

a travelling at a constant speed?

....

b accelerating?

....

c decelerating?

....

3 Look at the graph of a bus journey.

a Describe what is happening to the motion of the bus between these points:

O and A

....

A and B

....

B and C

....

C and D

....

b Calculate the acceleration of the bus between points D and E.

....

....

c Calculate the acceleration *and* deceleration of the bus between points B and C.

....

....

d Describe one way in which the graph is unrealistic.

....

4.5 Presenting results in tables and graphs

1 A student collected some data about other students in his class. First he timed how long it took 6 students to count to 50. Anyam took 16 seconds, Ejiro took 20 seconds, Mimi took 17 seconds, Ikenna took 17 seconds, Bayode took 14 seconds, and Iyo took 22 seconds.

a Present the results of the experiment in an appropriate table in the space below.

b On graph paper draw an appropriate graph.

c Explain why you have chosen to draw the type of graph that you have drawn.

....

....

2 Another student measured the heights of six other students and the time it took them to walk from one side of the room to the other. She wrote down the pairs of results like this: 1.35 metres and 7.2 seconds, 1.25 metres and 8.4 seconds, 1.10 metres and 9.1 seconds, 1.00 metres and 9.5 seconds, 1.40 metres and 8.1 seconds, 1.45 metres and 7.1 seconds.

a Present the results of the experiment in an appropriate table in the space below.

b On graph paper draw an appropriate graph.

c Explain why you have chosen to draw the type of graph that you have drawn.

....

....

3 You need to choose appropriate graphs for the data that you collect. Which graph should you plot in each of these experiments? Explain your answers.

a An experiment to measure the time it take to boil different volumes of water.

....

....

b An experiment to measure the extension of a spring as you add weights to it.

....

....

4.6 Asking scientific questions

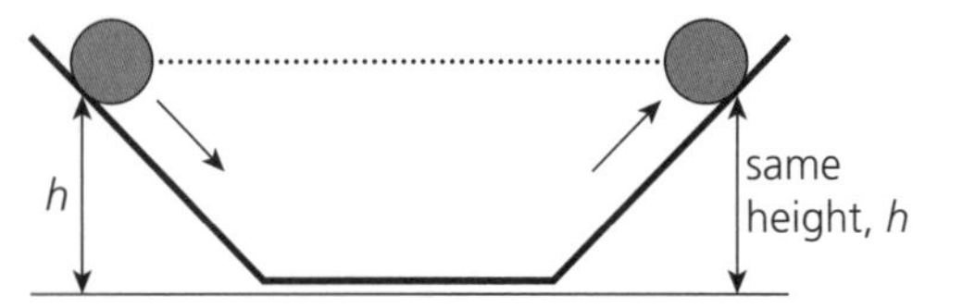

1 Here is a table of observations and predictions about rolling balls down slopes. The ball rolls down one slope, along a horizontal surface, and up another slope.

Complete the final column.

Observation	Prediction	Why this doesn't happen in real life
A ball rolling down a slope speeds up and a ball rolling up a slope slows down.	A ball on a horizontal slope does not speed up or slow down.	
A ball rolling up a shallower slope will roll further than up a steep one.	The ball will roll up a shallower slope until it reaches the same height.	
If you remove the second slope the ball will continue to move.	The ball will roll forever.	

2 Sometimes you can do experiments in science by using your imagination.

a What do we call this type of experiment?

....

b Lots of people think that heavier objects fall faster. This is because a very famous scientist called Aristotle thought this was the case. Imagine that Aristotle is correct. You have a heavy ball and light ball and you drop them. What will happen?

....

c Next you tie the balls together with a piece of string. You drop them again. What will happen?

....

d Finally you stick them together and drop them. What is the problem that Aristotle would have explaining this?

....

....

e How does this experiment show that heavier objects don't fall faster?

....

....

f Is it more convincing to see the experiment being done with equipment rather than in your imagination? Explain your answer.

....

....

E

a Explain why it takes a large force to get a massive object to stop.

b Put these objects in order of the force required to produce the acceleration.

c Why do many people think that you need a force to keep something moving?

	Mass (kg)	Acceleration (m/s^2)
A	100	5
B	100	20
C	50	5
D	200	20

5.1 Sound vibrations and energy transfer

1 Use the words and phrases from the box to complete the sentences below.
Use each word once, more than once, or not at all.

air	compressions	waves	solids	vibrating	rarefactions	moving	vacuum	radio

a All sound is produced by something that is ....
Sound can travel through solids, liquids, and gases but not through a ....

b Sound .... consist of .... (where the particles are close together) and .... (where the particles are further apart).

2 Here is a bar chart showing the speed of sound in different materials A, B, and C.

a Which material is a solid?

....

b Is material A a solid, a liquid, or a gas?

....

c How much faster, approximately, does sound travel in solids than in gases? Circle one of the answers below.

one thousand times faster **one hundred times faster** **ten times faster**

E When there is an earthquake, waves called seismic waves are produced.

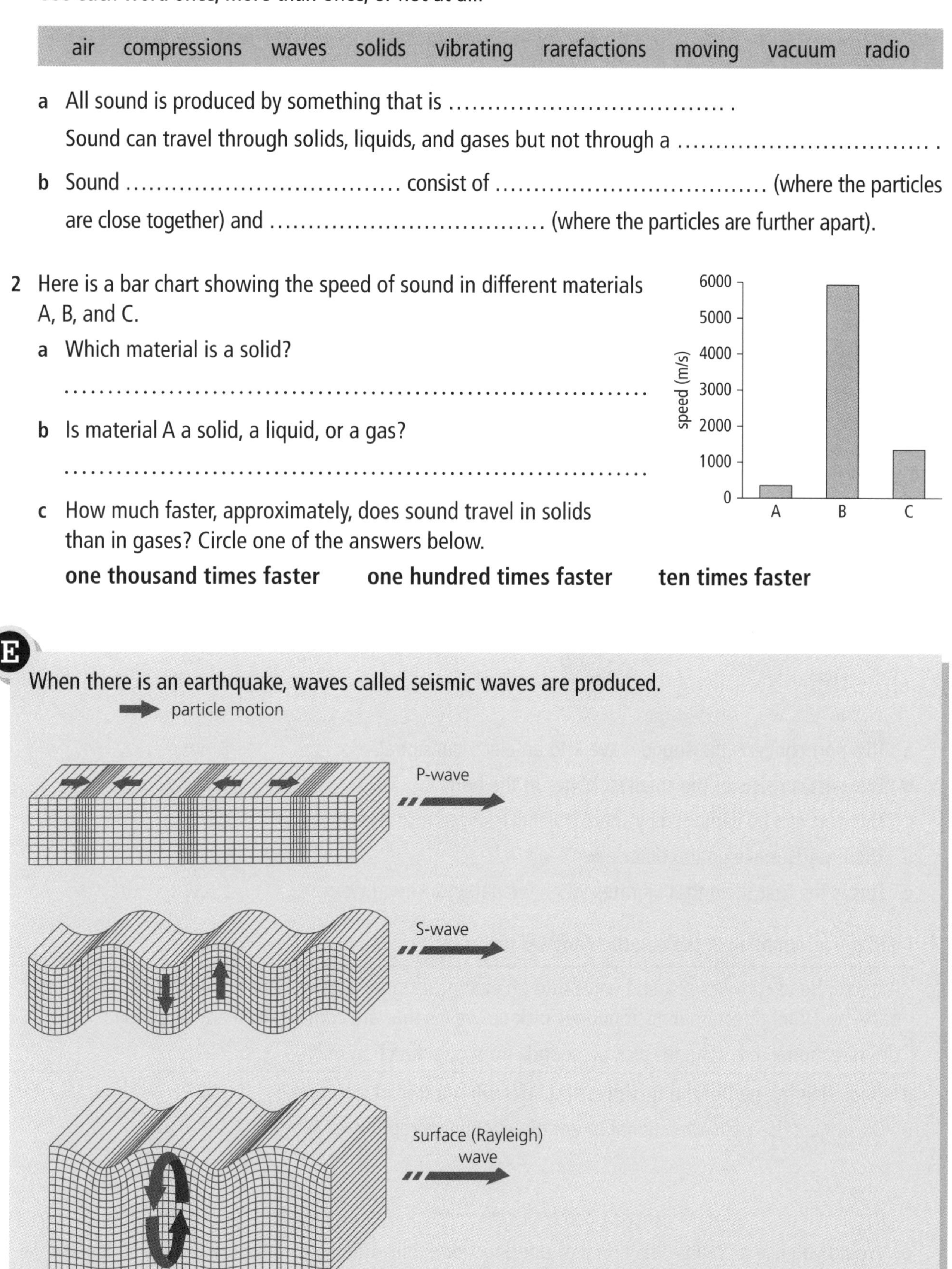

Decide which wave is a transverse wave and which is a longitudinal wave. Explain your answer.

5.2 Detecting sounds

1 Use the words and phrases from the box to complete the sentences below.
Use each word once, more than once, or not at all.

cochlea ossicles middle ear oval window inner ear eardrum auditory canal auditory nerve

a When you listen to music a sound wave travels along your ....
and makes your .... vibrate. This makes the small bones or
.... vibrate. These bones make up your ....
Your semicircular canals and .... make up your
.... The fluid inside the .... vibrates
when the vibration is passed on by the .... Signals produced by
sound-detecting cells travel down the .... to the brain.

b Which part of the outer ear is *not* in the list of parts of the ear above? What is its function?

....

....

2 Look at the diagram below. Identify the part that answers each of the questions below.

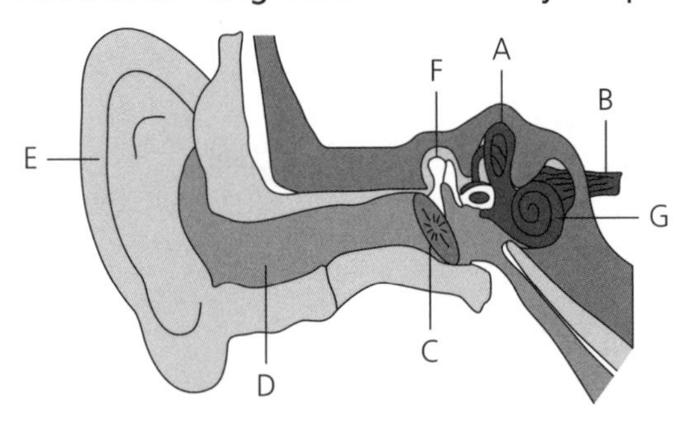

a This part converts the sound wave into an electrical signal. ....

b This part consists of the smallest bones in the body. ....

c This part can be damaged by sharp objects inserted into the ear. ....

d These parts make up the outer ear. ....

e This is the first thing that vibrates when we detect a sound wave. ....

3 Read the information in the box then answer the questions below.

> A microphone converts a sound wave into an electrical signal. Microphones have different pick-up patterns. Omni-directional microphones pick up waves that are coming from any direction. Uni-directional microphones pick up sounds from one direction only.

a Underline the part of the text that describes what a transducer does.

b Do singers use omni-directional or uni-directional microphones? Why?

....

....

c Would you use an omni-directional or uni-directional microphone to measure the sound levels in a classroom? Explain your answer.

....

....

1 A teacher is demonstrating how ear defenders work. She uses a loudspeaker to produce a sound and varies the loudness of the sound. Three students sit the same distance away from the loudspeaker. She asks each student to raise their hand when they can hear a sound. Here are the results:

Intensity (dB)	Ear defenders	Ear plugs	No ear plugs or defenders
5	hand not raised	hand not raised	hand not raised
10	hand not raised	hand not raised	hand raised
20	hand not raised	hand not raised	hand raised
30	hand not raised	hand raised	hand raised
40	hand not raised	hand raised	hand raised
50	hand raised	hand raised	hand raised
60	hand raised	hand raised	hand raised

a Are ear defenders or ear plugs better at reducing sound level? Explain your answer.

....

....

b Is the experiment a fair test? Explain your answer.

....

....

c Suggest one improvement to this experiment.

....

....

2 There are lots of ways of reducing the risk of damaging your hearing from loud sounds. Draw a line to connect the most appropriate method of reducing risk to each of the situations on the right.

Method	Situation
Wear ear defenders.	Working with loud machinery in a factory.
Move away from sound.	Living close to a noisy road.
Put something between you and the sound to absorb it.	Listening to loud music.
Reducing the length of time that you are exposed to the sound.	

E

Here is a graph that shows the maximum length of time you should spend exposed to different sound levels.

a Describe the link between sound level and time. Look at what happens to the time each time the sound level increases.

b Explain the link between sound level and time.

c A person buys a pair of ear defenders that reduces the sound level by 12 dB. How would that affect the length of time that they could wear them:

i If the noise level is 100 dB?

ii If the noise level is 109 dB?

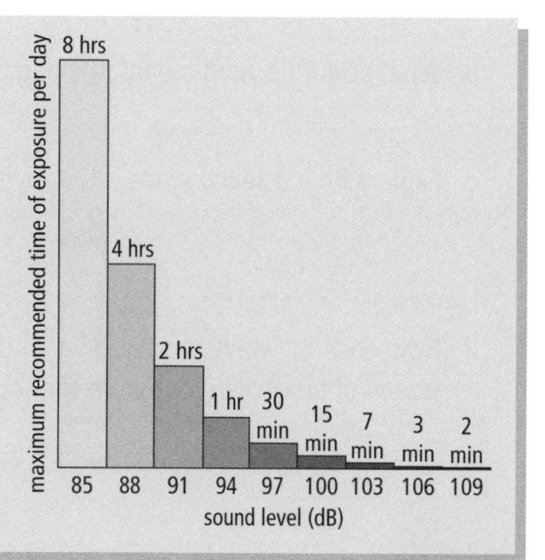

5.4 Loudness, amplitude, and oscilloscopes

1 A student has been learning about the properties of waves but is confused. This is what he writes:

a The wavelength of the sound is the distance from the middle to the top of a wave.

b An amplifier increases the wavelength of a wave.

c The frequency of a wave is the distance from the top of one wave to the top of the next wave.

d The amplitude of a wave is the number of waves per second.

Rewrite the sentences so that they are correct.

a ....

b ....

c ....

d ....

2 A student is holding a ruler as shown in the diagram. To make a sound she pushes the ruler down and lets it go.

a How is a sound produced?

....

b She holds a microphone near the ruler and shows the sound on an oscilloscope. Describe how the student can make the sound louder.

....

....

c Draw on the diagram to the right the wave that would be produced by the louder sound.

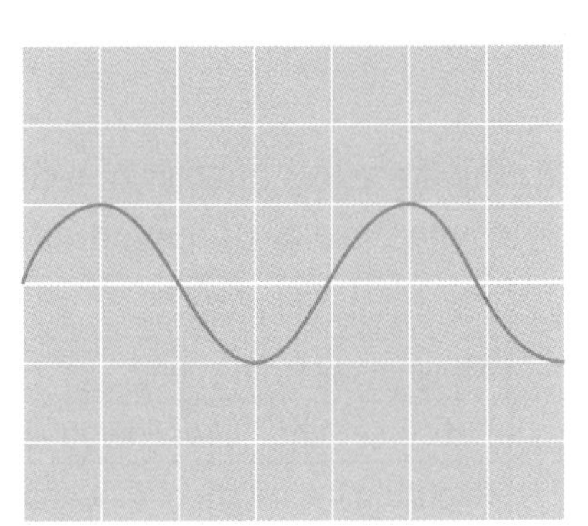

3 If you pluck a string it will make a sound. If you put a box underneath the string, like the box on a guitar, the sound will be louder.

a In terms of the properties of a wave, what is the difference between the sound wave that you hear with the box and the sound wave that you hear without the box?

....

b What scientific word means 'making the sound louder'?

....

c Explain how a sound wave is displayed on an oscilloscope.

....

....

d How does the wave produced when there *is* a box underneath the string look different on the screen of an oscilloscope from the sound when there is no box?

....

1 Circle the correct word or phrase in each **bold** pair in the sentences below.

The pitch of a sound depends on the **frequency / amplitude** of the sound wave.
A higher-pitched sound has a **higher / lower** frequency than a lower-pitched sound.
Humans hear sounds from about 20 Hz to 20 **million / thousand** Hz.
The frequency in Hz is the number of waves per **minute / second**.

2 A guitar has six strings and each is tuned so that it produces a different note. The table below shows the frequencies of the six strings.

Guitar string	Frequency (Hz)
1st	82
2nd	110
3rd	147
4th	196
5th	247
6th	330

a Explain what is meant by '82 Hz'.

....

b Which string will produce the highest note?

....

c Which string will produce the lowest note?

....

d The piano key 'A' produces a note of the same pitch as the 2nd string of the guitar. If you heard the same note played on the guitar and the piano you could tell the difference between them. How?

....

....

3 Here is a bar chart that shows the range of hearing of five people or animals.

a Which of A, B, C, D, and E is not a human?

....

b Which of A, B, C, D, and E is the youngest human?

....

c Which of A, B, C, D, and E is the oldest human?

....

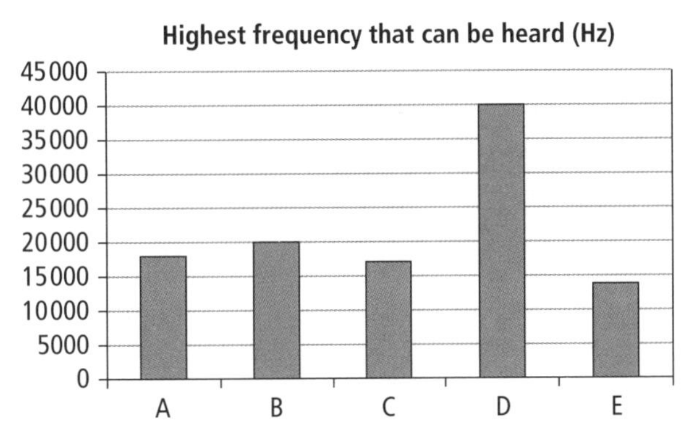

E

A student finds some information about the frequency of notes produced by different instruments that have pipes in them. The table below shows the length of the pipe and the frequency of the note produced.

Pipe length (m)	Frequency (Hz)
0.5	330
0.6	275
0.7	236
0.8	206
0.9	183
1.0	165

a On graph paper, plot the data as a graph.

b What is the link between the frequency and the length of the pipe?

c Another student says that the notes produced by longer pipes have a longer wavelength. Do you agree? Explain your answer.

5.6 Making simple calculations

1 Three students do calculations of the speed of sound using this information.

"A student is standing 100 metres from a wall. Her friend times how long it takes for her to clap in time with the echo 10 times. That time is 6 seconds. What is the speed of sound?"

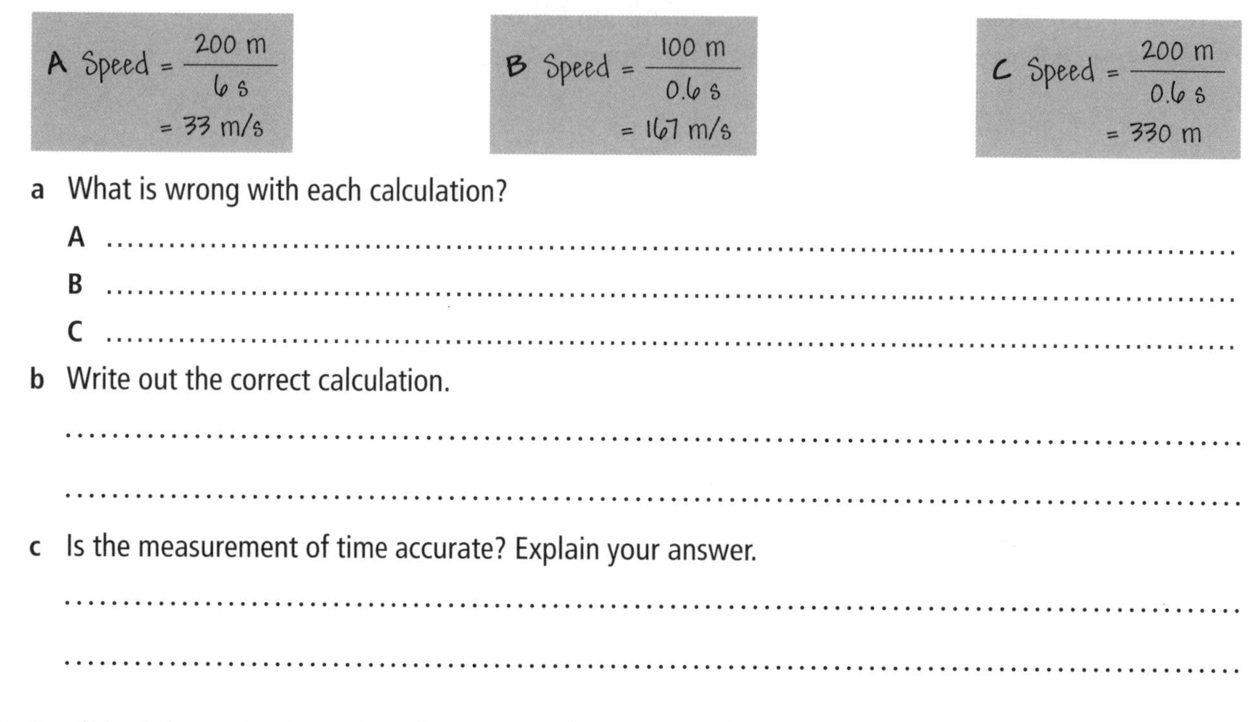

a What is wrong with each calculation?

A ....

B ....

C ....

b Write out the correct calculation.

....

....

c Is the measurement of time accurate? Explain your answer.

....

....

2 Read the information in the box then answer the questions below.

> For many years athletics races have been started with a starting pistol. This is a specially designed handgun that signals the start of the event. A sensor attached to the gun sends an electronic signal to the timing system to start the clock. It is also connected to a microphone that transmits a signal to loudspeakers located behind each of the runners. In races where the competitors have a staggered start, like the 200 metres, the loudspeakers are very important to make sure that the competitors all hear the gun at the same time.
>
> The starting pistol has now been replaced with an electronic system that sends a signal to each of the loudspeakers and starts the clock without making the sound of a gun.

a What can you say about the speed of the signal that is sent to loudspeakers compared with the speed of sound?

....

b Why are the loudspeakers even more important in staggered races than they are in races where competitors are in a line?

....

E

Mach numbers tell you how fast you are travelling compared with the speed of sound.

a How fast is something travelling if it is moving at Mach 3?

b What is the Mach number for a plane travelling at 495 m/s?

c How is an object moving if it has a Mach number less than 1?

5.7 Echoes

1 a A ship uses sonar to find a wrecked ship. Explain how sonar works.

....

....

The ship sends out pulses of ultrasound 5 seconds apart. Assume the speed of sound in water = 1500 m/s.

b What is the maximum depth the sonar can detect if the reflections are detected separately?

....

c Why does sonar use ultrasound rather than sound?

....

2 An ultrasound scan of a baby uses reflections to build up a picture of a baby.
Assume that the sound is travelling in water.
How long would it take for an ultrasound pulse to be reflected from a baby at a depth of 5 cm?

....

3 Dilip stands a distance from a cliff. His friend Ali stands between him and the cliff. Dilip fires a starting pistol and Ali hears a sound after 1 second, and another sound 3 seconds later.
Assume the speed of sound in air = 330 m/s.

a How far is Ali from Dilip?

....

b How far is Ali from the cliff?

....

c When would Dilip hear an echo from the cliff?

....

d How far is Dilip from the cliff?

....

4 The table below shows some famous wrecked ships and the depth at which they are found.

Name	Date lost	Where	Depth (m)
HMS *Titanic* (British passenger ship)	1912	North Atlantic	3800
MV *Bukoba* (Tanzanian ferry)	1996	Lake Victoria	25

a The *Titanic* was detected by sonar in September 1985, 73 years after it sank. Calculate the time for the echo from the *Titanic* to be received. Round it to the nearest 0.1.

....

....

b When a ship uses sonar it sends out pulses and receives the echoes. What is the shortest length of time between the pulses that you could use to find MV *Bukoba?* Explain your answer.

....

....

6.1 What is light?

1 Use the words and phrases from the box to complete the sentences below.
Use each word once, more than once, or not at all.

images	reaches	travels	shadows	transfers	reflections	travel	inverted	upright	straight

Light travels in .... lines. This explains how .... are formed on sunny days, how .... are formed in cameras, and why there are .... in mirrors. You cannot see something unless you have 'line of sight' and the light can travel from the object to your eye in a .... line. Light waves are one of the ways that energy from the Sun .... the Earth.

2 A student has been learning about shadows and how they are formed. She sets up a torch pointing at a wall. She puts a ball half-way between the torch and the wall. She measures the shadow, it has a width of 20 cm. She then moves the torch, ball, and screen to see what happens to the size of the shadow.

a Fill in the gaps in the table below using the phrase 'closer to' or 'away from'.

What the student did	Width of the shadow (cm)
Put the ball half-way between the torch and the shadow.	20
Moved the torch .... the ball.	30
Moved the torch .... the ball.	15
Moved the ball .... the wall.	10

b Would the shadow of the ball on the wall be sharp or would the edges be fuzzy? Explain your answer.

....

....

3 Here is a diagram showing some people and some buildings.

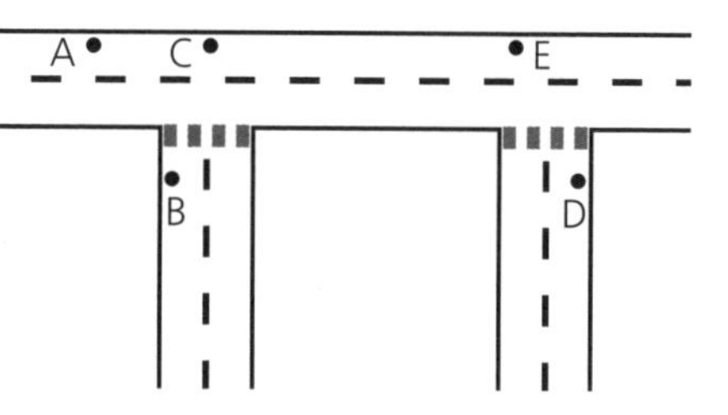

a Which pairs of people can see each other?

....

b Why can the other pairs of people *not* see each other?

....

....

E

A student looks at the image of a lamp on the screen of a pinhole camera.

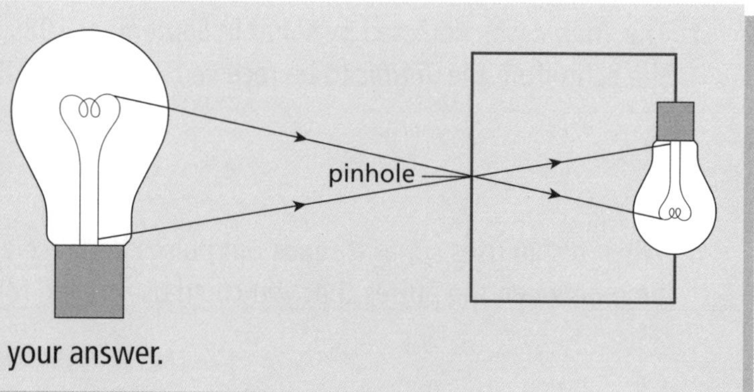

a Explain why he sees an inverted image.

b The student says that you can make the image the right way up by turning the camera upside down, but his friend says that you need to turn the lamp upside down. Who is right? Explain your answer.

1 Join boxes from each column to make three sentences about the way that light behaves.

You cannot see through …
You can see through …
Light can get through …

… transparent materials …
… translucent materials …
… opaque materials …

… but you cannot see through them.
… like glass.
… like concrete.

2 A student investigates the light-transmitting properties of some materials.

First he put the light meter on the other side of the material as shown in the diagram. He reads the light level on his light meter.

15 cm
power supply
ray box with single slit
material
light meter or sensor

a What is this a reading of? Is it a measure of how much light the material transmits or reflects?

....

Then he put the light meter on the same side as the ray box, pointing it at the material, and reads the light level.

b What is this a reading of? Is it a measure of how much light the material transmits or reflects?

....

3 A student shines a light with an intensity of 100 lux on different materials. She measures the reflected light intensity using a light meter.

Object	Reflected light intensity (lux)
A	25
B	85
C	65

a Which object is probably a mirror? Explain your answer.

....

b Which object is a light-coloured object? Explain your answer.

....

c Which object is a dark-coloured object? Explain your answer.

....

E

a Some parts of the eye transmit most of the light that hits them, and others absorb the light. Complete the table by putting a tick in one column for each part of the eye.

Part of the eye	Mainly transmitted	Mainly absorbed
pupil		
cornea		
rod cells		
lens		
cone cells		
eyelid		

6.3 The speed of light

1 In 1638, Italian scientist Galileo Galilei and his assistant performed an experiment to measure the speed of light. They stood on hills several kilometres apart. Galileo's assistant opened his lamp and when Galileo saw the light from it, he opened his own lamp.

lantern opened ...

...second lantern opened when light from first arrives!

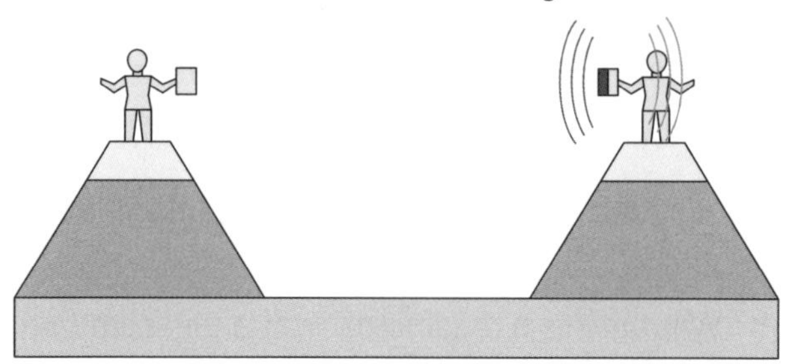

a The speed of light is 300 000 km/s. The distance between the two hills is 3 km. Choose the time delay that you would expect to see in this experiment. Circle the correct answer.

10 seconds

1 second

1/1000 of a second

1/100 000 of a second

b Could you detect this time delay with the human eye? Explain your answer.

....

c How far away would his assistant have to be to see a time delay of 1 second? What would be the problem with doing the experiment in this way?

....

....

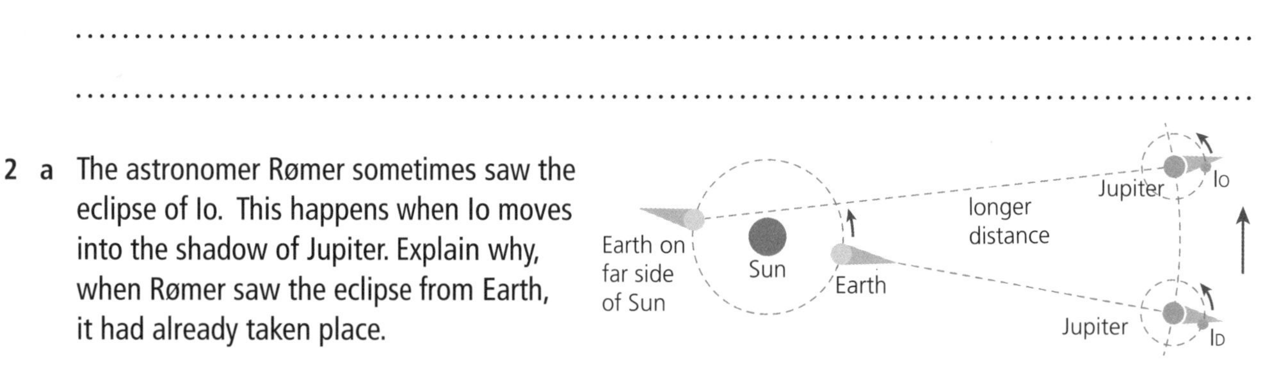

2 a The astronomer Rømer sometimes saw the eclipse of Io. This happens when Io moves into the shadow of Jupiter. Explain why, when Rømer saw the eclipse from Earth, it had already taken place.

....

b Sometimes Io is 6.2 million km away, and sometimes it is 9.4 million km away. Calculate these distances in light time.

....

c Describe the positions of the Earth, Sun, Jupiter, and Io when Io is the furthest from the Earth.

....

3 The robot Curiosity landed on Mars in August 2012. Communicating with robots on Mars is difficult, but even more so if you have to take into account the fact that light takes time to travel.

Look at the diagram of the inner planets of the Solar System on page 60 of the Student book. Explain why there is a range of values for the distance between Mars and Earth. (Mars and Earth travel at different speeds around the Sun.)

....

1 Write ***T*** next to the statements that are true. Write ***F*** next to the statements that are false. Then write the corrected versions of the statements that are false.

a The image that you see in a mirror is a real image.

b If you look in a mirror your image looks as if up and down are swapped over.

c The reflection of light that you see in a mirror is diffuse reflection.

d The image of an object in a mirror is the same size and shape as the object.

e Your mirror image appears closer to the mirror than you are.

Corrected versions of false statements:

....

....

2 When you look in a mirror you see your image.

a Complete the table.

Things that are *the same* about you and your mirror image	**Things that are *different* about you and your mirror image**

b You stand 50 cm in front of a mirror. What is the distance between you and your image?

....

E

You can achieve magic tricks with the reflection of light. In this trick you can make it appear that a candle is burning in a beaker of water.

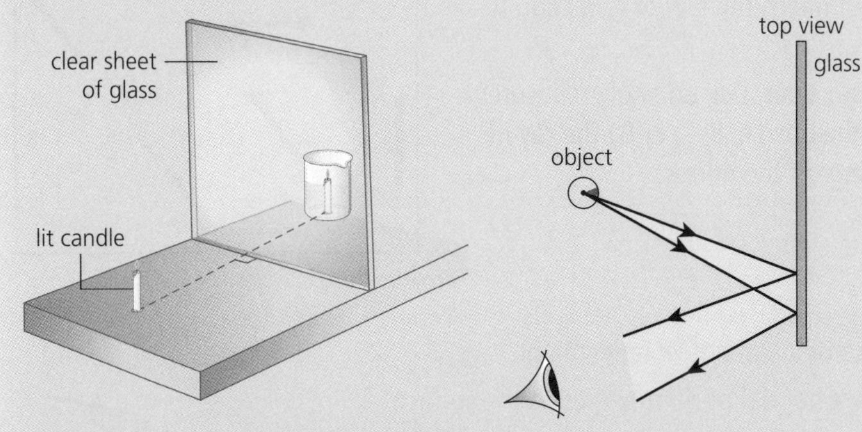

a What is behind the sheet of glass?

b Explain why it appears that the candle is burning underwater.

c Copy and complete the diagram to show how the image of the candle is formed in the glass.

d Explain how you can make it look as if an unlit candle is burning. Draw a diagram and explain what you have drawn.

1 Jamaal has done an experiment using a mirror to investigate how light is reflected.

He measures the angle of incidence and the angle of reflection. Here are his results plotted on a graph.

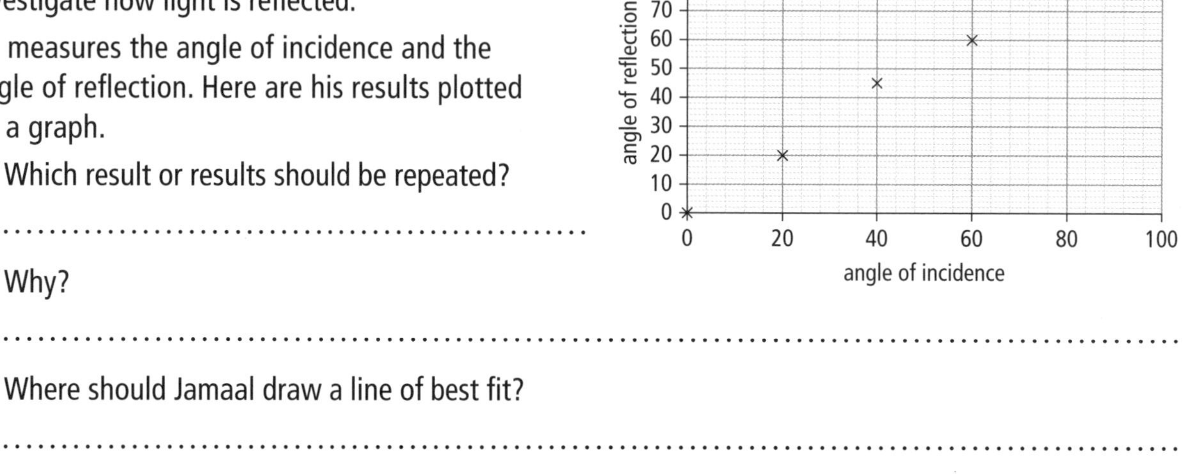

a Which result or results should be repeated?

....

b Why?

....

c Where should Jamaal draw a line of best fit?

....

d How do his results demonstrate the law of reflection? Explain your answer.

....

e He replaces the mirror with a white screen. The reflected ray is very faint. Why?

....

2 Look at the diagram.

Which of these statements is correct for the reflection of a light ray in a mirror? Put ***T*** in the final column if it is true and ***F*** if it is false.

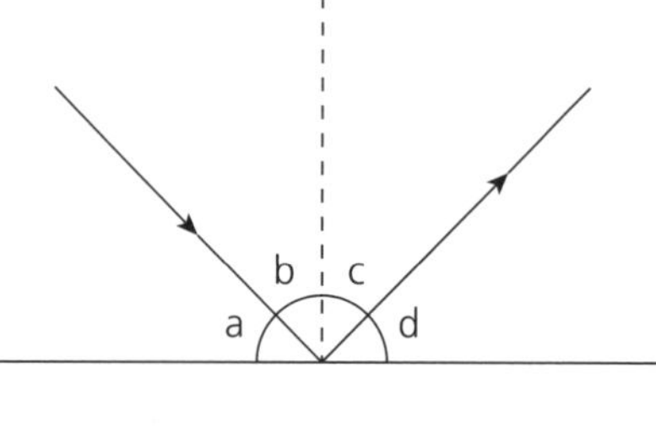

A	Angle a is always equal to angle b.	
B	Angle b is always equal to angle c.	
C	Angle c is always equal to angle d.	
D	Angle a is always equal to angle d.	
E	Angle a + angle b = 90 degrees.	

3 A student designs a puzzle that needs the law of reflection to be solved.

All of the mirrors reflect on both sides. Use an angle measurer to work out on which side of the box (A, B, C, or D) the ray of light will emerge after reflecting off the mirrors.

The ray will emerge on side: ....

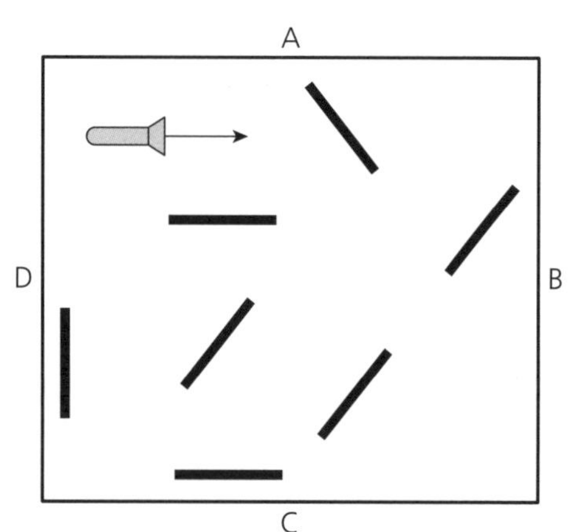

E

Uneven surfaces like the surface of a stone wall reflect light.

incident ray

uneven surface of stone wall

a On the diagram above draw a normal at right angles to the surface where the ray hits it.

b Draw an arrow to show the reflected ray.

c Draw another incident ray parallel to the ray drawn above that hits a *different* part of the wall.

d Draw another normal and reflected ray to show how this ray is reflected.

e Explain why you cannot see your face in a stone wall.

1 A student is finding out what happens when light goes from water into air. She puts a coin in the bottom of the cup and puts the cup on a table. She walks away until she cannot see the coin at the bottom of the cup.

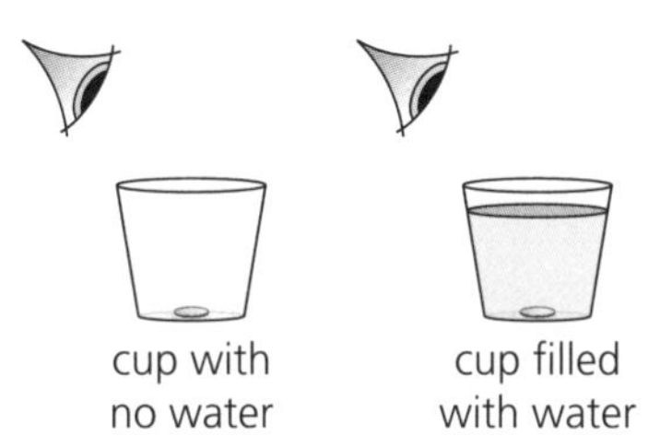

a Use what you know about how we see things to explain why she cannot see the coin when there is no water in the cup.

....

....

b Her friend now pours water into the cup while she is looking at the coin. The coin appears. Use what you know about how we see things to explain why she can see the coin when there is water in the cup.

....

....

c Complete the diagram of the cup filled with water showing rays from the coin in each case.

2 Mirages are very common in deserts. You see what appears to be a sheet of water a short distance ahead of you. You don't need to go to a desert to see a mirage. On a hot day you often see the same effect on a road. The road in front of you may appear wet or shiny but, however far you travel, you never reach the water because it is an optical illusion.

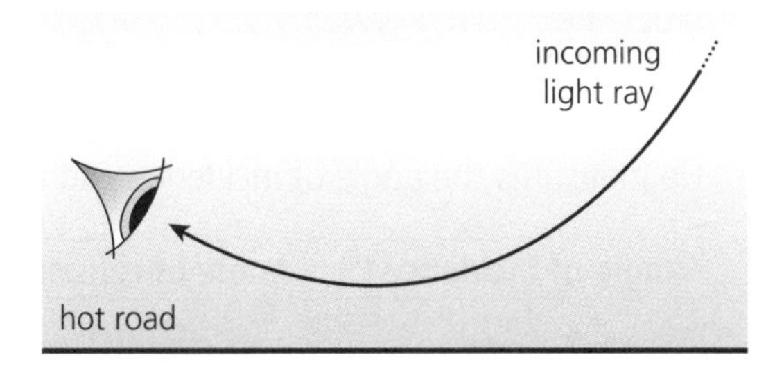

What you are seeing is an image of the sky, which looks like water to us.

a The brain works by assuming that light travels in straight lines. Draw a line to show where the light appears to come from.

b As the light travels toward the road is it moving into air that is denser or less dense?

....

c Use your answer to part **b** to explain why the ray bends as it does.

....

....

E

Refractive index tells us how much light is slowed down by a medium such as air or glass.

a Complete the table:

Material	Speed of light in a vacuum (km/s)	Speed of light in the material (km/s)	Refractive index
vacuum	300 000		1
glass	300 000	200 000	
alcohol	300 000		1.36
salt	300 000	190 000	

b Why does refractive index not have a unit?

c Is it possible to have a material with a refractive index of less than 1.0? Explain your answer.

1 Complete the following sentences using the words from the list.
Each word may be used once, more than once, or not at all.

denser direction refraction incidence incident parallel perpendicular quickly refracted slowly straight

Light is .... when it passes through a glass block. The angle of

.... (*i*) is the angle between the normal and the .... ray.

The angle of .... (*r*) is the angle between the normal and the

.... ray. The ray of light changes direction when it enters the glass

block because glass is .... than air and the light travels more

.... The ray of light changes direction when it leaves the glass block because

light travels more .... in air than in glass. The rays entering and leaving the

block are ....

2 Deepak has been studying what happens when light goes into glass.
He measures the angle of incidence and angle of refraction.

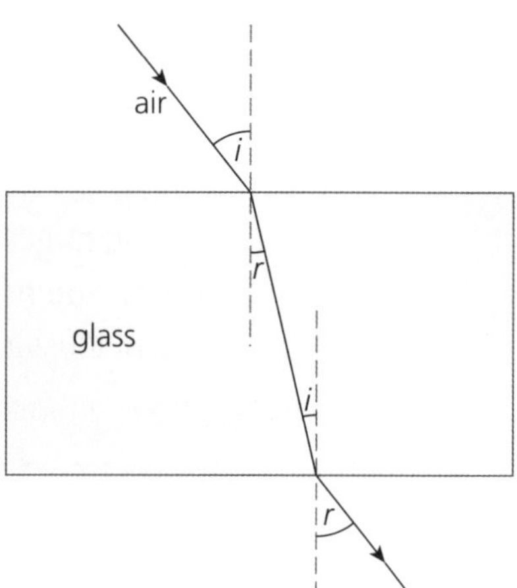

Angle of incidence (°)	Angle of refraction (°)
10	7
20	13
30	37
40	25
50	31

a What is the angle of refraction if the angle of incidence is zero?

....

b Which result or results does not fit the pattern?

....

c Deepak replaces the block with one that has a lower refractive index.
Write a prediction that he could make about what will happen to the angle of refraction.

....

....

E

Light travels in straight lines but can travel around corners in an optical fibre.

a Explain how you can see around corners but still have the light travelling in straight lines. Draw a diagram and explain it.

b Look at the speed of light in different materials. Write a list of the materials with the material that refracts *most* first and the material that refracts least last.

Material	Speed of light in the material (million km/s)
air	300
diamond	125
glass	200
plastic	187
water	225

6.8 Dispersion

1 Write ***T*** next to the statements that are true. Write ***F*** next to the statements that are false. Then write corrected versions of the statements that are false.

a The spectrum is made up of nine colours.

b Violet is refracted more than red.

c Raindrops change white light into coloured light.

d Light is reflected as it goes through a prism.

Corrected versions of false statements:

....

....

2 The diagram shows a ray of white light being dispersed by a prism.

a On the diagram label **R** where you would see red light in the spectrum and **V** where you would see violet light.

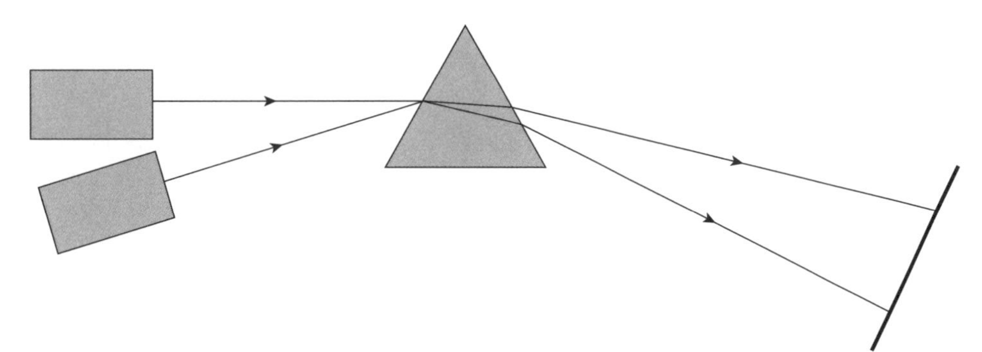

b Draw a circle around the points on the diagram that shows light being refracted.

c You can add another prism to recombine the light and produce white light. On the diagram draw where you would put a second prism.

3 Rainbows can be formed wherever white light interacts with drops of water. This can happen when it rains or over waterfalls. Put the following statements in order to explain how a rainbow is formed.

A Different colours are refracted by different amounts as the light enters the drops.

B The white light enters raindrops and is refracted.

C The Sun emits white light, which is made up of all the different colours of light.

D As the light leaves the raindrops it is refracted again spreading the colours out even more.

E You have to stand with the light behind you to see the rainbow because the light has been reflected from the back of the raindrop.

F All the colours of light are reflected from the inside surface of the drop.

E Different colours are refracted by different amounts, and this produces dispersion. Explain why. You must include these words in your explanation: wavelength, frequency, speed.

1 Use the words and phrases from the box to complete the sentences below.
Use each word once, more than once, or not at all.

red	cyan	reflects	magenta	green	absorbs	blue
yellow	secondary	primary	filter	transparent	transmits	

There are three .... colours of light: ...., red, and ....

You can combine these colours to make .... colours: ....,

......................, and .... All the other colours of the spectrum can be made

by mixing different amounts of the .... colours. That is why colour television or

computer screens are made of segments that emit ...., ..., and

.... light. A .... is made of a material that .... one

colour of light and .... the rest of the colours. It is only .... to

one colour of light.

2 a What do filters do to light? Complete the diagram to show what happens when light goes through filters.

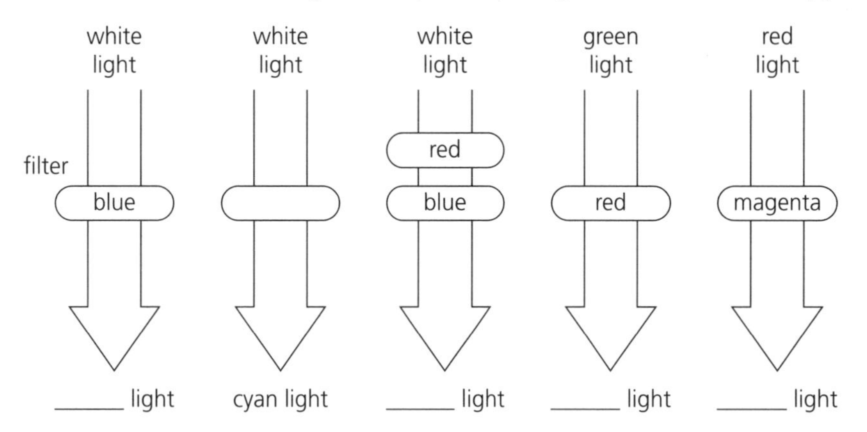

b Is the light transmitted by a filter brighter than, less bright than, or the same brightness as the light that hits it? Explain your answer.

....

....

E

A student makes a spectrum using a ray of white light and a prism. He looks at the colours on a screen. Describe *and explain* what he would see on the screen in each of the situations below.

a He puts a green filter between the ray box and the prism.

b He puts a red filter between the prism and the screen.

c He puts a blue filter between the ray box and the prism and a red filter between the prism and the screen.

6.10 Presenting conclusions: more on colour

1 Circle the correct word or phrase in each **bold** pair in the sentences below.

We see objects the colour they appear to be because in white light they **transmit / reflect** some colours of light, and **absorb / transmit** all the rest. A blue object will **reflect / absorb** blue light so that the object appears to be blue. It will **reflect / absorb** all the other colours of light. Objects that appear black **reflect / absorb** all colours of light and objects that appear white **reflect / absorb** all colours of light.

2 The diagram shows rays of light reflecting from paper of different colours. Label the arrow to show the colour of light that the paper appears.

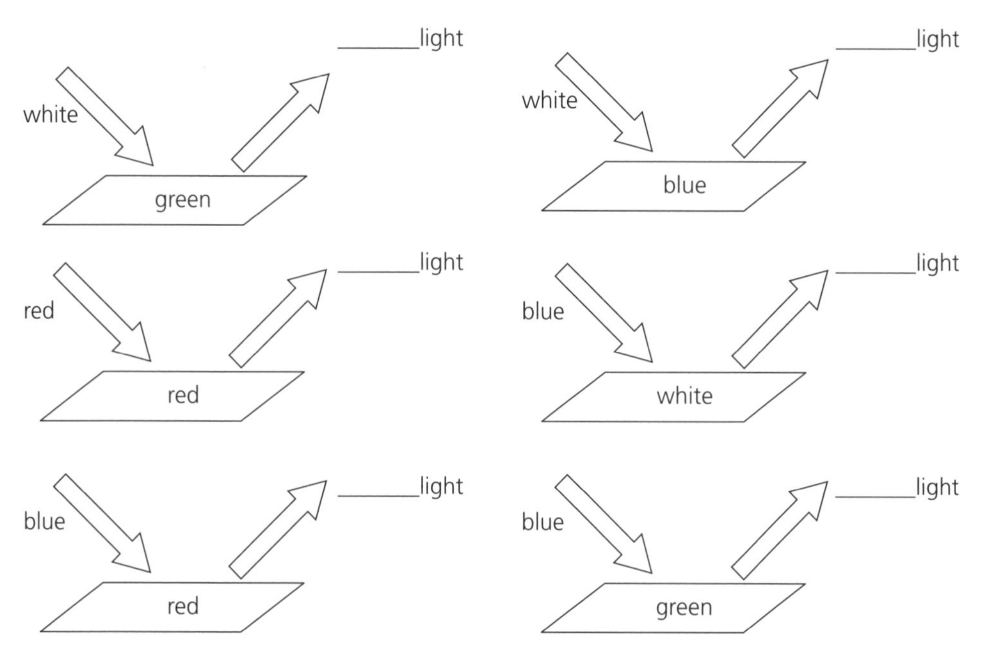

3 A child is playing with some coloured building blocks. His brother uses a torch and different filters to shine light of different colours on the blocks. Use the words red, green, blue or black to complete the gaps in the table below.

Colour of block	Colour of light hitting object	Colour block appears
blue	red	....
blue	magenta	....
red	....	black
green	....	green

E

a Which light-sensitive cells are being used to make the image?

b A student is holding out a pen at arm's length. He looks at the pen out of the corner of his eye and it looks black and white. He cannot tell what colour it is.

c He looks directly at the pen and can now see that it is red. Which light-sensitive cells are being used to make the image now?

d Are there rods or cones in sections A in the retina?

e Are there rods or cones in sections B in the retina?

f Explain your answers to parts **c** and **d**.

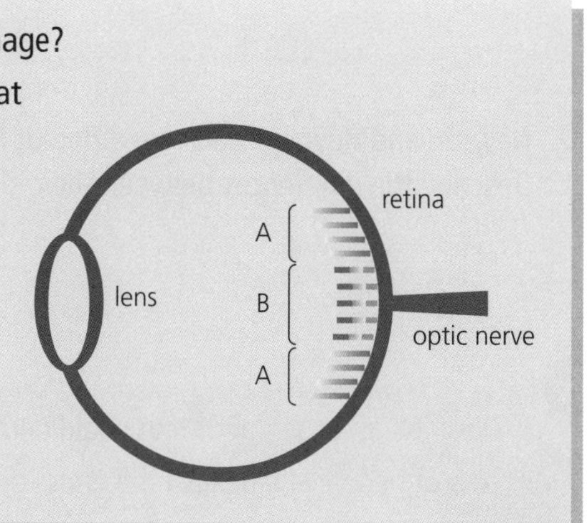

6.11 Asking scientific questions

1 Read the information in the box then answer the questions below.

We know that we see things because light is reflected from them into our eyes. However, this is not the only idea that people have had about why we see things.

The first idea was put forward 2500 years ago. This idea was that something comes out of your eye that enables you to see things. It is called the 'emission theory'. This is because light is emitted from your eye and you see the object when the light hits it.

People thought that if you closed your eyes at night then opened them the light from your eyes travels immediately to the stars so you can see them.

The second idea was that we see things because something from the object comes into our eyes. This was called the 'intromission theory'.

People thought that the light would take time to travel from the object into your eye.

a Describe the difference between the intromission and the emission theories.

....

....

b Here are some observations. Suggest whether each one supports or undermines the emission theory, the intromission theory, or both. Explain your answers.

i Some animals, like cats, have eyes that seem to light up at night.

....

....

ii Someone with weak eyes does not see something better when someone with strong eyes looks at it.

....

....

iii You can only see things in your 'line of sight'.

....

....

2 Newton and Huygens had very different ideas about light.
Describe the differences between their ideas.

....

....

E

a Describe an experiment that could demonstrate that the emission theory cannot be true.

b Lots of people still believe the emission theory. Suggest one reason why.

1 Lasers have changed the way that music is recorded, stored, or transferred.

a Describe how lasers are used to record music onto a CD.

....

....

b Describe how lasers are used to playback music from a CD.

....

....

c This is a model of how the CD works. The pits represent a string of 1's and 0's, which is a binary number. The flat sections between pits are called 'lands'. A signal made of 1's and 0's is a digital signal.

land

pit

1 0 1 0

If there is a change in the height then the CD will record a 1. If there is no change it will record a 0. The diagram above shows the number 1010.

This can be converted into a normal, or decimal, number. 1010 is the same as the number 10.

Decimal	1	2	3	4	5	6	7	8	9	10
Binary	0001	0010	0011	0100	0101	0110	0111	1000	1001	1010

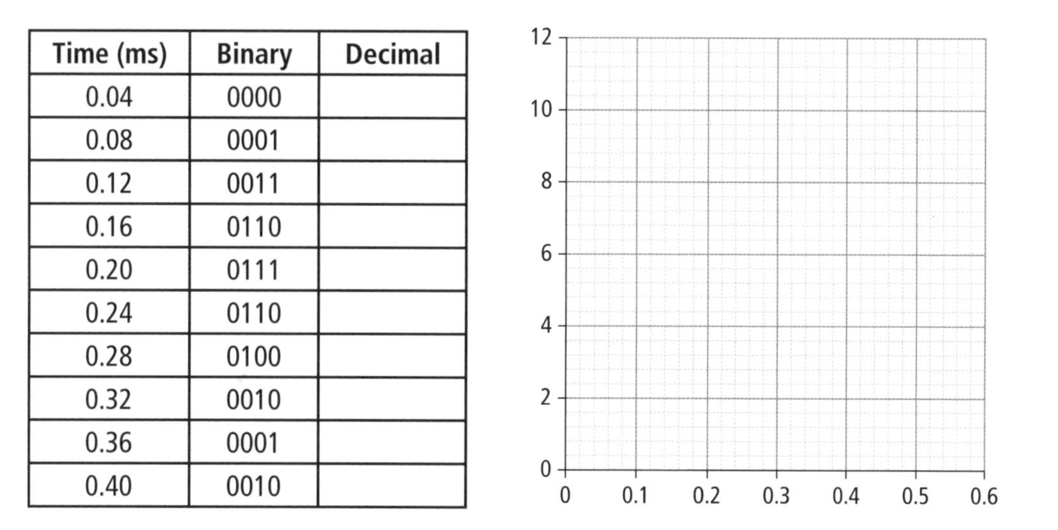

Time (ms)	Binary	Decimal
0.04	0000	
0.08	0001	
0.12	0011	
0.16	0110	
0.20	0111	
0.24	0110	
0.28	0100	
0.32	0010	
0.36	0001	
0.40	0010	

i Convert the numbers in the table into decimals.

ii Plot the numbers on the graph and join them up to make a wave.

d You can record better quality sound by using more 1's and 0's. What impact would that have on the number of tracks on the CD?

....

e The distance between the pits on the CD depends on the wavelength of the light of the laser. If the wavelength is smaller, the pits can be closer together and they can still be read.

i Will a red or a blue laser store more songs on a CD?

....

ii Explain your answer.

....

1 Use the words and phrases from the box to complete the sentences below.
Use each word once, more than once, or not at all.

iron	south	attracted to	copper	north	repelled by

Magnetic materials such as ...., nickel, or steel can be magnetised.

When you move a magnet near a magnetic material the material will be ....

the magnet, but it will never be .... the magnet. ....

nails are .... a magnet because each one becomes a magnetised.

If the pole that is touching the nail is a north pole, then the end of the nail touching the magnet has

become a .... pole.

2 A child has been given a magnetic fishing game. The description says:
"This is a game based on magnetism that uses rods to pick up wooden fish from the box."

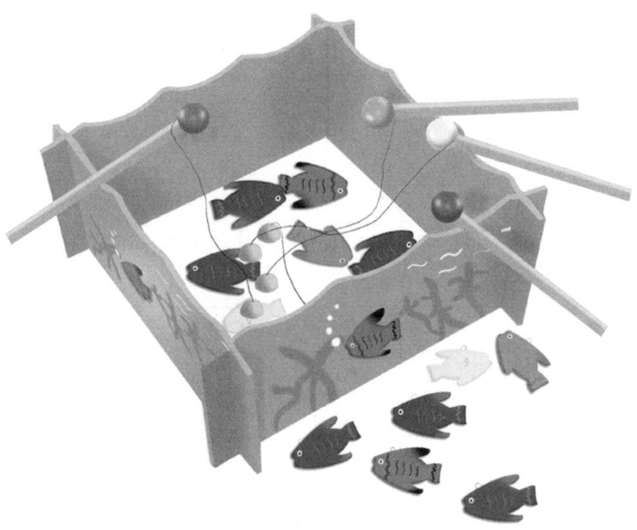

a How do you know that the 'fish' are not made of wood?

....

b Here are possibilities for how the rods and fish have been made. Put a tick (✓) in the blank column if you think that the game would work.

The ball at the end of the rod is a magnet and the fish are made of a magnetic material.	
The ball at the end of the rod is made from a magnetic material and the fish are made from a magnetic material.	
The ball at the end of the rod is a magnet and the fish are magnets.	

c Explain your answer to **b**.

....

E

Here is a drawing of someone starting to use a magnet to magnetise a rod of metal.

a On the diagram:
- label the magnet
- label the rod
- draw domains inside the magnet
- draw domains inside the unmagnetised rod.

b The rod is now magnetised. Is A the north pole or the south pole?

N S

A B

1 Magnets can be made in lots of different shapes. On each of these diagrams on the left *either* add the names of the poles *or* add the field lines to complete the patterns.

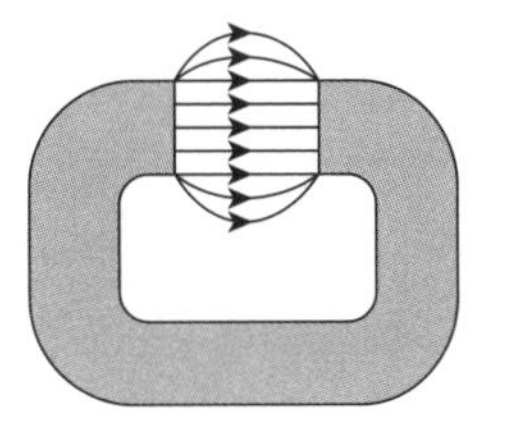

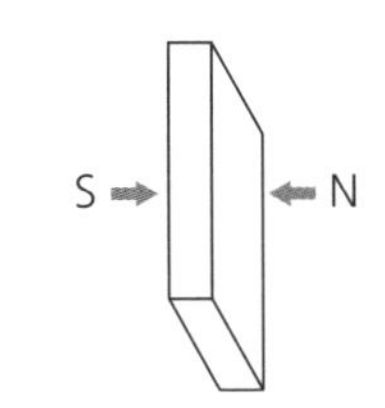

2 Draw the magnets on these diagram to show what magnet, or combination of magnets, would create each pattern. Label the north and south poles.

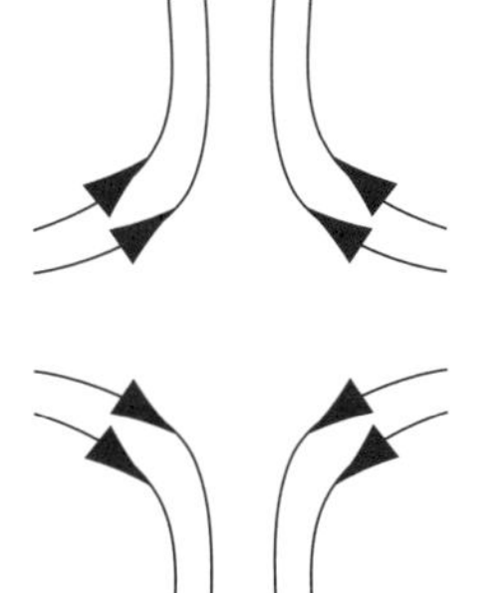

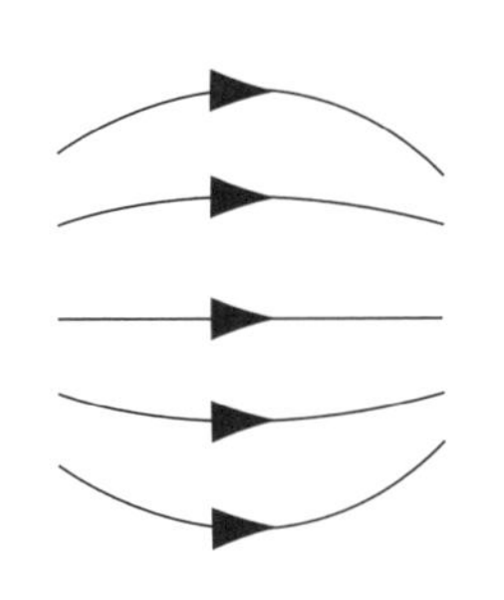

3 The Earth's magnetic field enables us to navigate with compasses. In this picture a compass is shown next to a magnet. You take the compass away from the magnet and hold it in your hand. Imagine you are standing on the equator and follow the instructions below. Write down what you would see.

N S

a You face north and look down at the compass in your hand. Is the north end of the compass needle pointing in front of you, behind you, to your left or to your right?

....

b You turn to face west and look down at the compass again. Is the north end of the compass needle pointing in front of you, behind you, to your left or to your right?

....

c You turn to face south and look down at the compass again. Is the north end of the compass needle pointing in front of you, behind you, to your left or to your right?

....

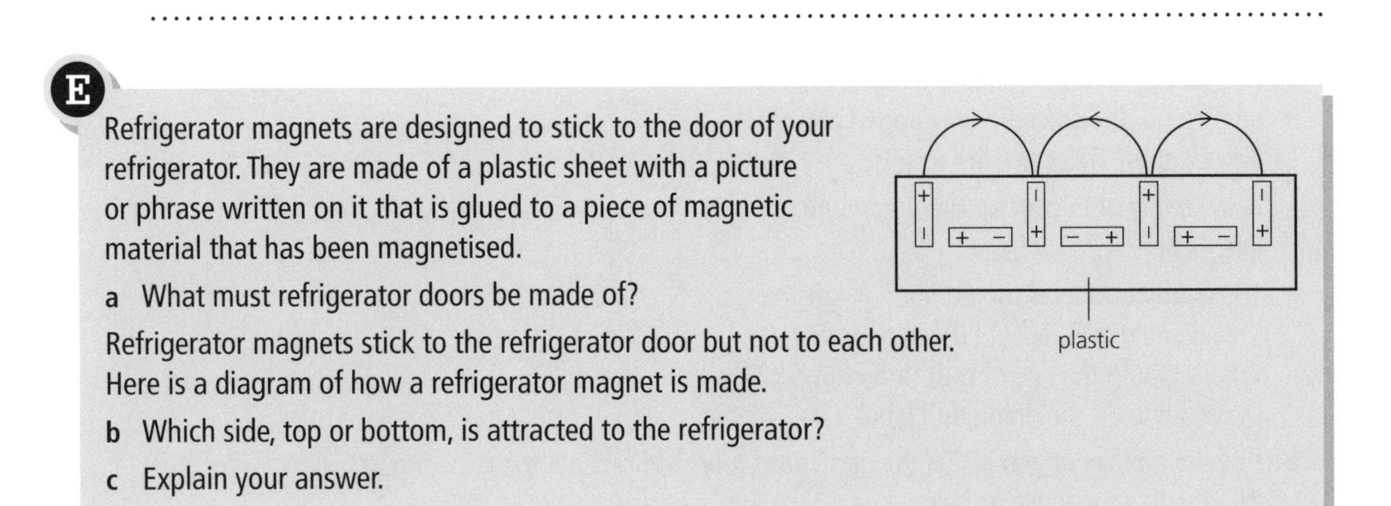

E

Refrigerator magnets are designed to stick to the door of your refrigerator. They are made of a plastic sheet with a picture or phrase written on it that is glued to a piece of magnetic material that has been magnetised.

a What must refrigerator doors be made of?

Refrigerator magnets stick to the refrigerator door but not to each other. Here is a diagram of how a refrigerator magnet is made.

b Which side, top or bottom, is attracted to the refrigerator?

c Explain your answer.

1 Write ***T*** next to the statements that are true. Write ***F*** next to the statements that are false.

a The magnetic field around an electromagnet is the same as the magnetic field around a straight wire.

b Replacing the iron core of an electromagnet with an aluminium core will make it stronger.

c Unlike permanent magnets, electromagnets can be turned off.

d If you change the direction that the current is flowing in an electromagnet, the poles of the electromagnet will swap over.

e An electromagnet with more coils will pick up fewer paper clips.

f Permanent magnets are always stronger than electromagnets.

2 A student investigates the magnetic field around a wire.
She puts a plotting compass on the card.

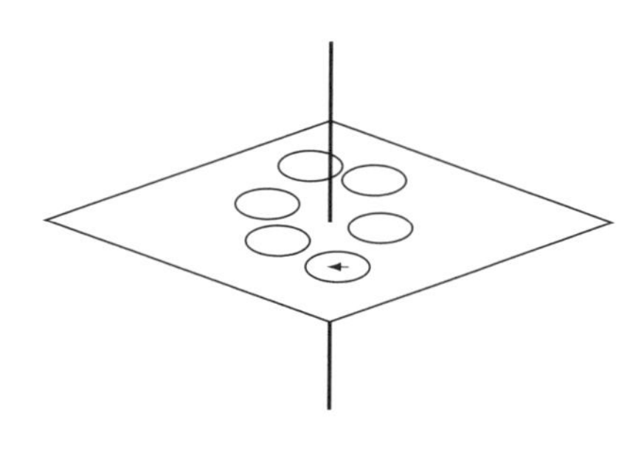

a Add arrows to the other circles to show the direction that the compass needle will point if more compasses were placed on the card.

b What would happen to the compass needles if she reversed the battery connections?

....

3 A student has copper wire and steel wire, nails made of iron or steel, and a battery.

a Explain how the student could make an electromagnet using this equipment.

....

....

....

b Does the choice of nail matter? Explain your answer.

....

....

E

A student changes the current flowing in an electromagnet and measures the paper clips that it can pick up. The solenoid is wrapped around a cardboard tube. These are the results.

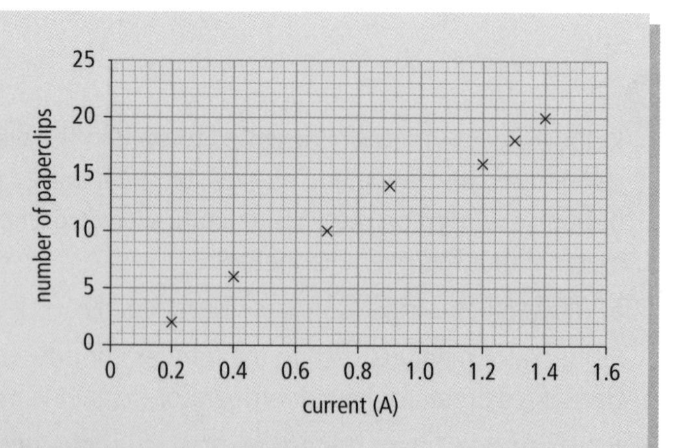

a Draw a line of best fit on the graph and label it A.

b The student doubles the number of coils on the solenoid and repeats the experiment. Where would the line of best fit be now? Draw a line on the graph and label it B.

c He puts a piece of iron inside the cardboard tube and repeats the experiment. Will the line be above or below line A? Explain your answer.

7.4 Identifying and controlling variables

1 A table of results for an investigation into the effect of changing the number of coils on the strength of an electromagnet is below.

Number of coils	Number of paper clips	Number of paper clips	Number of paper clips	Average number of paper clips
5	5	6	7	
10	9	8	7	
15	11	13	12	
20	17	18	19	
25	21	22	26	
30	27	25	26	
35	30	34	32	

a Complete the table by calculating the average number of paper clips for each number of coils.

b Write a list of the variables that need to be controlled in this experiment.

....

....

....

2 A student changes the number of coils, voltage, type of core, and type of wire in an electromagnet investigation.

Experiment	Number of coils	Voltage of the battery (V)	Type of core	Type of wire	Number of paper clips
A	10	3	iron	copper	4
B	20	6	iron	copper	12
C	30	3	iron	steel	3
D	10	3	aluminium	copper	0
E	30	3	iron	copper	12
F	20	3	iron	copper	8

a Which experiments show the effect of changing the number of coils? ....

b Which experiments show the effect of changing the voltage? ....

c Which experiments show the effect of changing the type of core? ....

d Which experiments show the effect of changing the type of wire? ....

3 Sinta uses large paper clips to measure the strength of her electromagnet. Maharani uses small paper clips instead.

a Will the size of the paper clip affect the strength of the magnet?

....

b Suggest one advantage of using small paper clips in this investigation.

....

c Suggest one advantage of using large paper clips in this investigation.

....

7.5 Using electromagnets

1 Electromagnets can be very useful. Here is a circuit that contains an electromagnet.

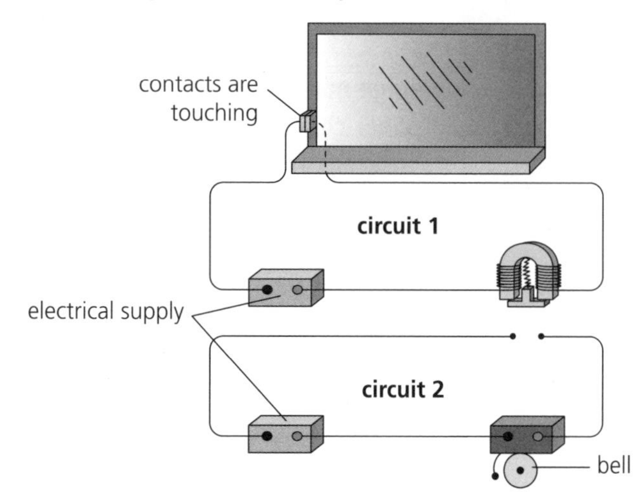

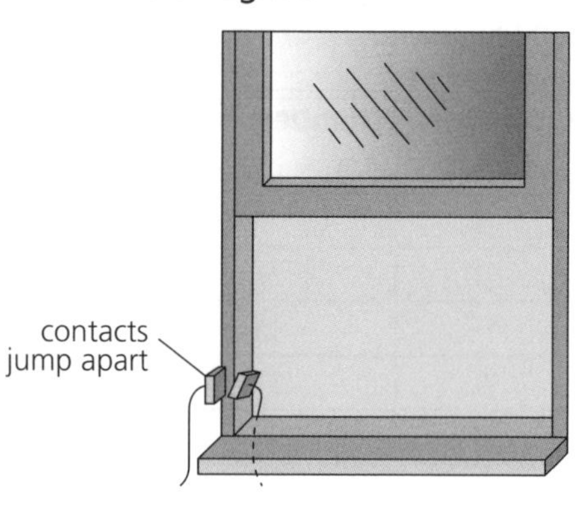

It is designed to set off an alarm if the window is opened. Explain how the circuit works.

....

....

....

2 A circuit breaker is a device that protects an electric circuit from overheating and possibly starting a fire or damaging electrical equipment. This is a simplified diagram of a circuit breaker.

The diagram shows the position of the circuit breaker when a **normal** current is flowing in the circuit. The same current that is flowing in the circuit also flows through the electromagnet. The armature is pushing the contacts together.

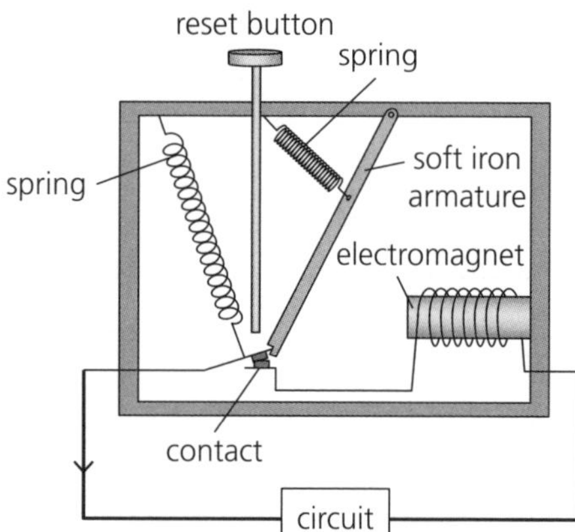

a Explain how the circuit breaker stops the current flowing when it gets too big.

....

....

....

b Explain how the circuit breaker can be reset so that the current can flow in the circuit again.

....

....

....

3 How can electromagnets be used in hospitals?

....

....

....

....

1 Circle the correct word or phrase in each **bold** pair in the sentences below.

Pressure is a measure of how much **force / mass** is exerted over a certain **volume / area**. If the force applied to a certain area is bigger the pressure will be **smaller / bigger**. If you increase the area over which a force is exerted the pressure will be **smaller / bigger**. You can calculate pressure using the equation **force divided by area / area divided by force**. You can measure pressure in **newtons / pascals**.

2 Use this equation to work out the pressure:

a A box weighs 50 N. Its base has an area of 5 m^2. What pressure does it exert?

....

....

b Another identical box is placed on top of the first box. What is the pressure now?

....

....

c A snowboarder weighs 300 N. The area of the snowboard is 0.5 m^2.
What is the pressure on the snow?

....

....

3 A student is investigating the pressure that a block of wood is exerting on the floor. The weight of the wood is 20 N. The block is 10 cm long, 5 cm wide, and 2 cm high.

a What is the *biggest* pressure that the block can exert on the floor?

....

....

b What is the *smallest* pressure that the block can exert on the floor?

....

....

E

Copy the table. Use the equation above to complete the table. Remember to use the correct units for pressure and area.

Force (N)	Area	Pressure
20	4 cm^2	
60		1.5 N/m^2
	12	0.05 N/m^2
75		15 N/cm^2

Here are some situations where pressure is important.
For each situation say how the ideas of force, pressure, and area can be applied.

The head of a drawing pin has a large area but the tip has a small area.	It hurts to walk on sharp stones but not on round ones.	You make a deeper footprint in mud than on dry ground.
A farmer hammers a pole into the ground that has been sharpened.	You could use a plank of wood to rescue someone from quicksand.	If you hold a heavy bag with a narrow handle it can hurt your hand.
If you build a house in a swampy area you need very wide foundations.	Hockey boots have studs.	Animals that live in muddy areas have big feet.

E

The length of a knife is 15 cm, and when it is sharp its width is 0.05 cm.

a Calculate the area of the knife blade.

b The cook pushes with a force of 15 N on the blade. Calculate the pressure.

Over time the blade becomes blunter. Its width is now 0.15 cm.

c Calculate the area of the knife blade.

d The cook pushes with a force of 15 N on the blade. Calculate the pressure now.

e How much force must the cook use to cut with the blunt knife now?

1 Write ***T*** next to the statements that are true. Write ***F*** next to the statements that are false. Then write corrected versions of the statements that are false.

a The pressure in a liquid decreases with depth.

b The upthrust on an object is larger when it is deeper in a pool.

c The bottom of a dam is thinner than the top of a dam.

d The pressure is bigger at the bottom of a lake because of the weight of water above it.

Corrected versions of false statements:

....

....

2 a A teacher has a round flask with holes in it. As she pushes down the water comes out of the holes in the way shown in picture A. Explain why.

....

b She repeats the experiment with a flask with no holes. What will happen? Why?

....

....

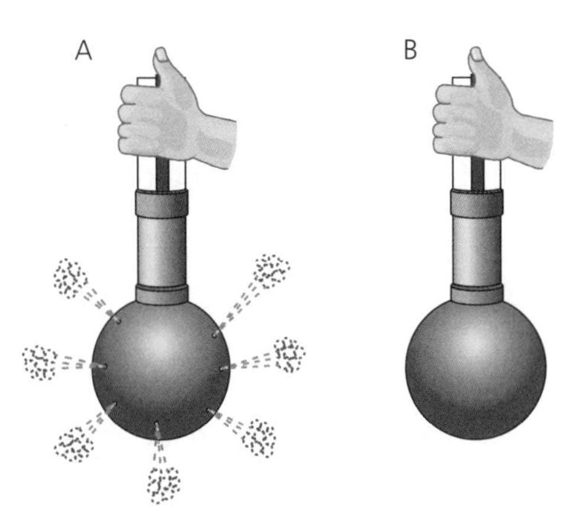

3 Divers have to be very careful when they come up from deep water. Use the words below to explain why. Use each word once, more than once, or not at all.

increase oxygen nitrogen carbon dioxide decompresses bends low high decrease

a As a diver comes up the pressure will .... The .... which dissolved in the blood will come back out.

b If the diver .... too quickly the .... leaves the blood so quickly it will bubble and fizz like .... from a can of fizzy drink.

c A diver can go into a special chamber to avoid getting the '........................'. The pressure in the chamber is controlled to change slowly from a .... pressure to a .... pressure.

E

A diver takes readings of the pressure at different places in a swimming pool.

The reading on the pressure gauge when it is at A is 12 kPa.

a How many N/m^2 are there in 12 kPa?

b How does a pressure gauge work?

c He moves the pressure gauge to B and C. Choose the correct pressure readings from the list in each place. Explain your answers.

12 kPa 24 kPa 6 kPa 18 kPa

8.4 Using pressure in liquids

1 Here is a diagram that shows how a car brake system works.

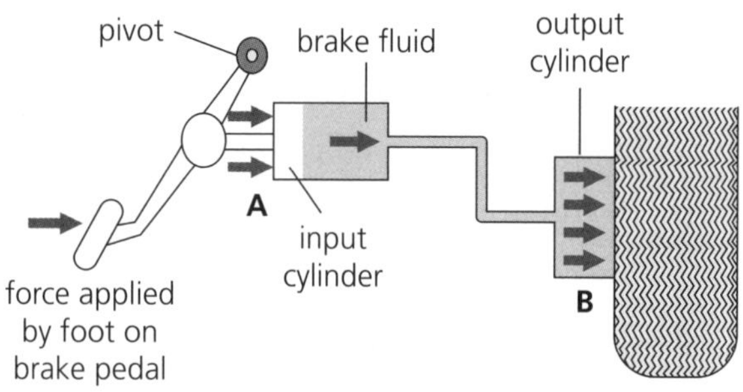

a A car braking system is a hydraulic system. What does hydraulic mean?

....

b Use the words and phrases from the box to complete the sentences below.
Use each word once, more than once, or not at all.

incompressible	input	bigger	output	compressible	oil	air	smaller

When the driver pushes on the brake pedal the .... cylinder A is pushed in.

This produces a .... in the liquid. Liquids are .... so this pressure is transmitted through the liquid to the .... cylinder. It is very important that there is no .... in the brake fluid. The force need to stop the car is much .... than a person could produce by themselves.

c The driver applies a force of 300 N to the brake pedal. The area of the input cylinder is 2 cm^2.
The area of the output cylinder is 80 cm^2.
Calculate the pressure in the liquid.

....

....

d Explain why a braking system is a *force multiplier.*

....

e A student thinks that you are getting 'something for nothing'.
Compare the distance moved by the output cylinder with the distance moved by the input cylinder.
Complete this sentence:

A small force moves through a .... distance on the input cylinder and a big force moves through a .... distance on the output cylinder.

E

Another hydraulic machine is a hydraulic jack.
It is also a force multiplier.
The jack is used to lift cars. Valve 1 opens when the mechanic pushes down. Valve 2 opens when he pulls the handle back up. Explain why the mechanic has to pump the handle lots of times to lift the car. (*Hint:* look at your answer to part **e** of the previous question).

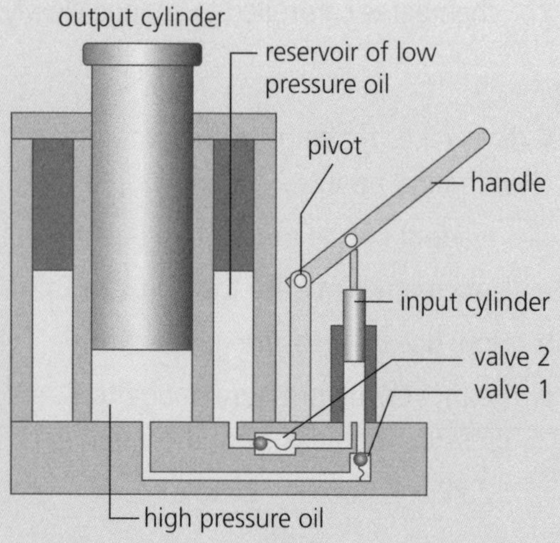

8.5 Pressure in gases

1 Match the beginning of each sentence with the correct end of the sentence. Connect the boxes with lines.

The particles in a gas are …
If you compress a gas …
Gas pressure is produced …
The particles in a liquid are …
If you try to compress a liquid …

… close together.
… spread out.
… you cannot compress it.
… when molecules collide with a container's walls.
… the particles are closer together.

2 In the diagram to the right where is the atmospheric pressure greatest?

Explain your answer.

....

....

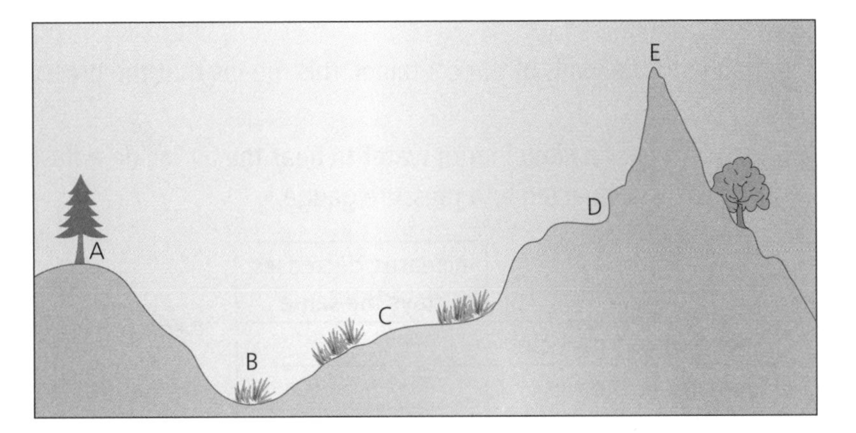

3 A teacher demonstrates air pressure to the class.

She takes a soft drinks can and puts a little bit of water in it. Then she heats it up until the water is boiling. The students watch as the steam pushes air out of the can.

The teacher then quickly takes the can, turns it over and dips the open end in a bowl of water. A short time later there is a loud bang and the can collapses.

a When the teacher cools the can some of the steam turns back to water. What happens to the air pressure inside the can?

....

b Why? (Use ideas about collisions in your answer.)

....

....

c Explain why the can collapses.

....

....

E

a What happens when you drink through a straw? When you suck the straw. There is air inside the straw that is pushing down on the liquid. The atmosphere is pushing down on the liquid. Use these ideas to explain why the liquid moves up through the straw.

b There is a limit to the length of straw that you can use to suck the liquid up. It is almost impossible to drink through a straw that is more than 3 m long. Why?

8.6 Pressure, volume, and temperature in gases

1 Use the words and phrases from the box to complete the sentences below.
Use each word once, more than once, or not at all.

more | less | increases | decreases | collide with | stick to | faster | slower | different | volume | pressure

The molecules in a gas are moving in .... directions. The pressure in a gas is produced when the gas molecules .... the walls of the container.

When the gas is compressed into a smaller ...., molecules collide .... often with the walls of the container. This means that the pressure ....

2 A student uses a container of water to heat the air inside a flask.
The flask is connected to a pressure gauge.

	Increases, decreases, or stays the same
Speed of gas particles	
Pressure in the flask	
Mass of particles	
Volume of gas	

Complete the table to show what happens to the gas in the flask as the temperature is increased. Use the words **increases**, **decreases**, or **stays the same**.

3 Here are three cylinders containing the *same* number of molecules of oxygen. The pressure in each cylinder is the same.

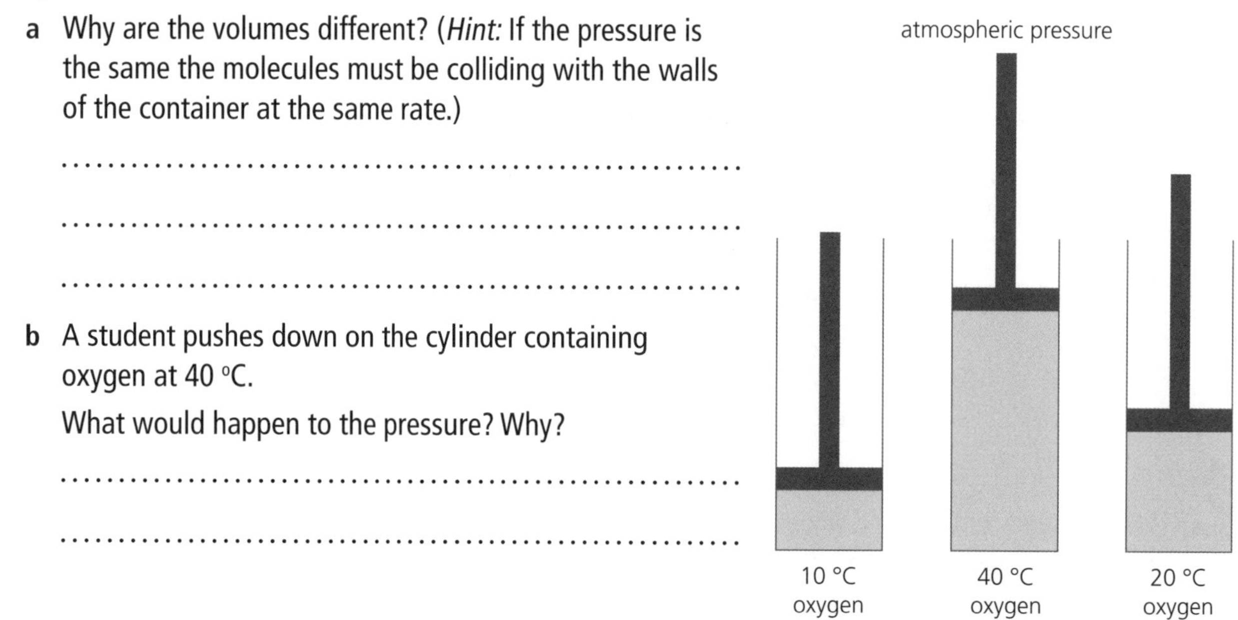

a Why are the volumes different? (*Hint:* If the pressure is the same the molecules must be colliding with the walls of the container at the same rate.)

....

....

....

b A student pushes down on the cylinder containing oxygen at 40 °C.
What would happen to the pressure? Why?

....

....

8.7 Preliminary work

1 A student has learned that thermometers work because liquids expand when you heat them up. She wants to do an experiment to investigate how liquids expand.

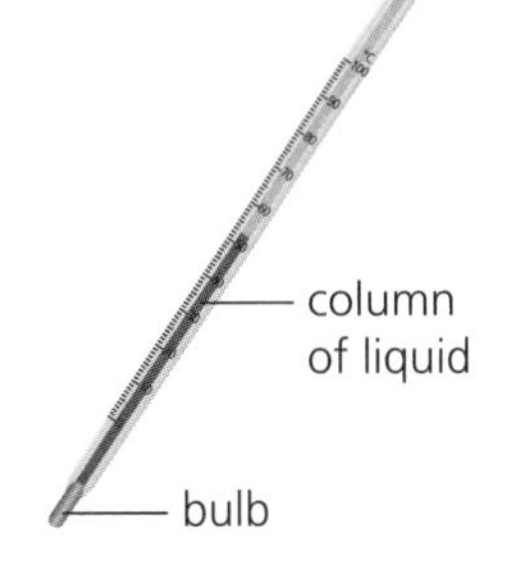

Her idea is to put a tube in a beaker of liquid, heat the liquid, and see how much the liquid rises up the tube.

Here is some of the preliminary work that she did for her investigation.
In each box write down how that preliminary work would help her to plan her experiment.

Preliminary work	What the preliminary work was for
She tried lots of different sized diameters.	
She tried very hot and very cold water.	
She tried lots of different ways of measuring how far the liquid went up the tube.	
She looked up the hazards of using different kinds of liquids.	
She tried different liquids.	
She worked out how many times she would need to repeat her experiment.	

E You are thinking of doing an experiment about the effect of the temperature on the volume of a gas. Describe the preliminary work that you could do and how it would help you to plan your investigation.

1 Here are some measuring instruments and the quantities that they measure.

a Match the measuring instrument with the quantity that it measures. Draw lines between them.

balance
measuring cylinder
ruler

liquid volume
mass
volume of an regular solid

b Describe one thing that is important to do to measure liquid volume accurately.

....

....

2 Complete the table. Use the correct units of density. Round your answer to 2 decimal places.

Material	Mass	Volume	Density
helium	4 kg	23 m^3	
salt	110 g	50 cm^3	
stone	26.5 g	10 cm^3	
oxygen	42 kg	30 m^3	

3 Here are some statements which describe one method to find the density of an irregular object.

A Fill a measuring cylinder or beaker with water.

B Find the mass of the object.

C Subtract the volume without the solid from the volume with the solid.

D Divide the mass by the volume to get the density.

E Measure the volume of water in the beaker or measuring cylinder.

F Put in the solid and measure the volume again.

a Put the statements in the correct order using the letters.

b Why are there two possible orders for the letters?

....

c Write down the other order.

E

One of the densest material in the Universe is the material in a neutron star. Very massive stars can form neutron stars when they collapse. The mass of 1 cm^3 is about the same as the population of all the people on Earth. There are 7 billion (thousand million) people on Earth and their mass is about 70 kg each.

a Calculate the density of a neutron star.

b An elephant has a mass of 3500 kg. 1 cm^3 of a neutron star has the mass of how many elephants?

c One of the least dense materials on Earth is aerogel. It is a very, very light foam. Some people call it 'solid smoke'. It has a density of about 1 kg/m^3. The rooms in a house have a volume of 75 m^3 and the mass of air in them is 90 kg. Is aerogel more or less dense than air?

8.9 Explaining density

1 Circle the correct word or phrase in each **bold** pair in the paragraph below.

The density of a solid is **bigger / smaller** than the density of a liquid. This is because the particles in a liquid **are / are not** arranged in a regular pattern. The density of a gas is much **bigger / smaller** than the density of a liquid because the particles in a gas **are / are not** much further apart than they are in a liquid.

2 Submarines and fish can change their density, so that they can move up and down in the oceans. A submarine uses a tank around the outside of the inner tank where people live. The outer tank can be flooded or can have air pumped into it.

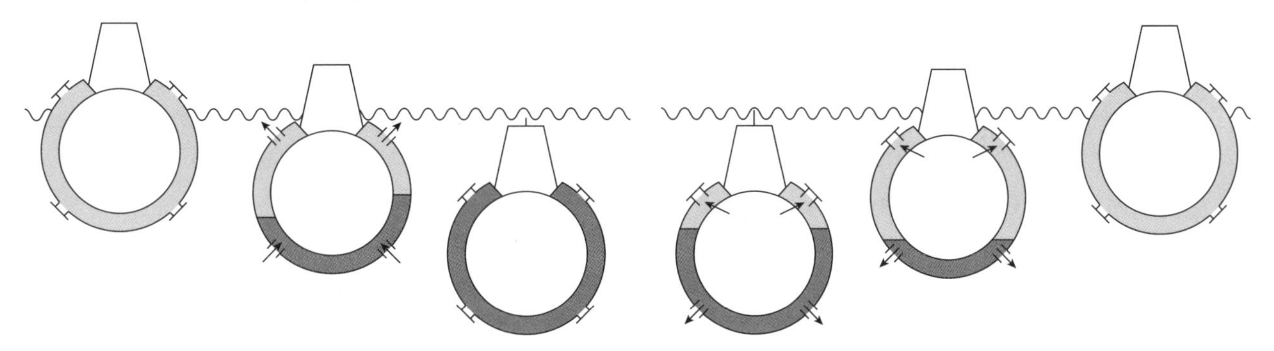

a Explain how these tanks are used to submerge the submarine. Use ideas about density in your answer.

....

....

b Explain how these tanks are used to surface the submarine. Use ideas about density in your answer.

....

....

3 Blocks of four different materials are in a tank of water. All the blocks have the same volume.

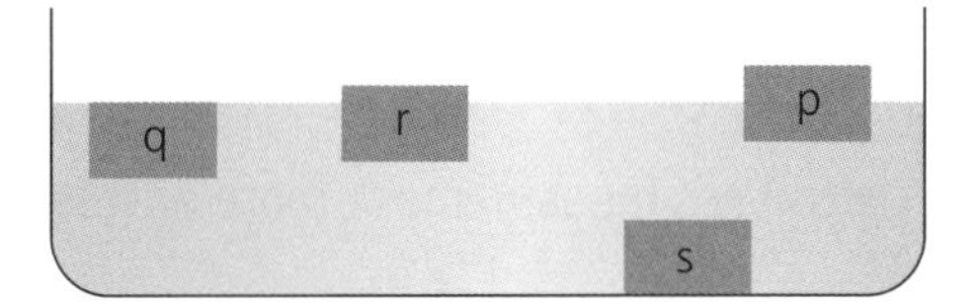

Write ***T*** next to the statements that are true.
Write ***F*** next to the statements that are false.

a Q is the least dense and S is the most dense.

b S is the densest and R is the second densest.

c Q is less dense than P.

d R is more dense than P.

e None of the blocks is less dense than water.

E

The mass of a submarine, including the air inside the inner and outer tanks, is 8 000 000 kg. The volume of the submarine is 10 000m^3.

When the outer tanks are full of water the total mass of the water in the tanks is 6 500 000 kg.

Density = mass / volume. The density of water is 1000 kg/m^3.

Use this information to explain why the submarine cannot float when the outer tanks are full but can when they are full of air.

8.10 Questions, evidence, and explanations

1 Read the information in the box, then answer the questions below.

> You have learned about measuring the densities of liquids, such as water, or solids, such as wood. You can measure the volume and mass of the objects and calculate the density using the equation:
>
> $$\text{density} = \frac{\text{mass}}{\text{volume}}$$
>
> How do you measure the density of something much bigger, like a planet?
>
> The first person to measure the density of the Earth was a scientist called Henry Cavendish in 1798. He was very shy and worked at home. He spent many, many hours doing experiments and hardly ever saw anyone else. He did lots of different experiments in chemistry and in physics.
>
> He used Newton's law of gravitation to make measurements to calculate the mass of the Earth. He used another scientist's measurement of the volume of the Earth to calculate the density. The result that he got was 5.48 g/cm^3. The measurement that scientists have made today is 5.513 g/cm^3
>
> Many centuries later scientists wanted to know the density of things in space, such as stars and black holes. They used observations from telescopes all over the world to put lots of measurements together to work out those densities. They worked out that the density of the Sun is 16.220 g/cm^3

a In what way or ways was Henry Cavendish like Al-Biruni?

....

....

b In what way or ways was Henry Cavendish not like Al-Biruni?

....

....

c Give one reason why the measurement of the density of the Earth that scientists have measured today is different from Cavendish's measurement.

....

d Is today's measurement more or less *precise* than Cavendish's measurement?

....

e Is today's measurement more or less *accurate* than Cavendish's measurement? Explain your answer.

....

....

f How were the methods for working out the density of the Sun different from the way that Cavendish worked?

....

....

E

a How much more dense is the Sun than the Earth? Choose one of these and circle it:

ten times **three times** **twice**

b A student says that the reason that we get better measurements of things like density is because we have better measuring instruments. Do you agree? Explain your answer.

1 Use the words and phrases from the box to complete the sentences below.
Use each word once, more than once, or not at all.

multiplier	effort	load	pivot	bigger	machine	force

A lever is a simple .... that acts as a force .... The lever turns about a .... The force that you apply is called the .... and the force that is produced is called the .... To use a small force to lift a big weight you need a lever where distance between the .... and the .... is bigger than the distance between the .... and the ....

2 a What is meant by a 'turning force'?

....

b Where is the pivot in the diagram of the spanner? Label it P.

c Which letter A, B, or C is the pivot in the diagram of the door handle?

....

d Where would you apply the force to the door handle?

....

e What would you notice if you applied it at the other labelled point?

....

....

3 It is much easier to open a tin of paint with a screwdriver than with your fingers.

a Label the pivot in the diagram.

b Draw an arrow to show the distance from the pivot to the load L.

c Draw an arrow to show the distance from the pivot to the effort E.

d Use what you have done to explain why the force you need to open the can with a screwdriver is very small.

....

....

8.12 Calculating moments

1 Which of these statements correctly state the law of moments? There may be more than one. Write ***T*** next to the statements that are true. Write ***F*** next to the statements that are false.

a The clockwise and anticlockwise moments are the same.

b The clockwise and anticlockwise moments add up to zero.

c The force × the distance on the left of the pivot = the force × the distance on the right of the pivot.

d The total clockwise moments = the total anticlockwise moments.

2 A student is playing a balancing game.
The weight of each monkey is 0.1 N.
The distance between holes and the pivot and the first hole on each side is 4 cm.

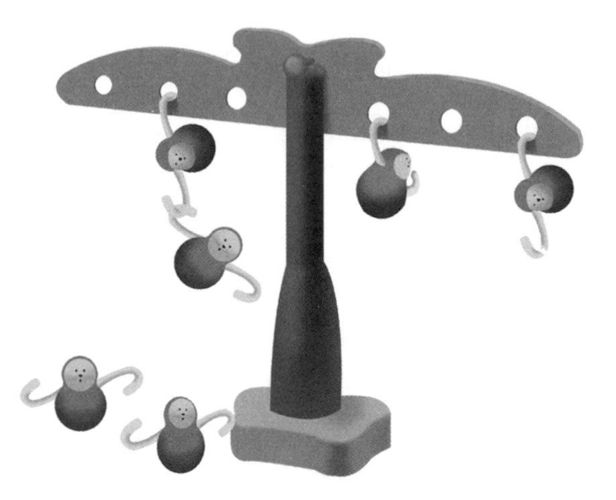

a Calculate the total clockwise moment.

....

b Calculate the total anticlockwise moment.

....

c Is the picture correct? Explain your answer.

....

....

d The student hangs another monkey on the left-hand side under the other two. Would it be possible for her to add another monkey on the right-hand side to balance? Explain your answer.

....

3 Priya and Tom are on a see-saw pivoted at the centre. Priya, who weighs 400 N, sits 2 m from the centre on the left. She is balanced by Tom who weighs 500 N.

a How far is Tom from the centre?

....

....

b Draw a diagram of the arrangement in the box below.

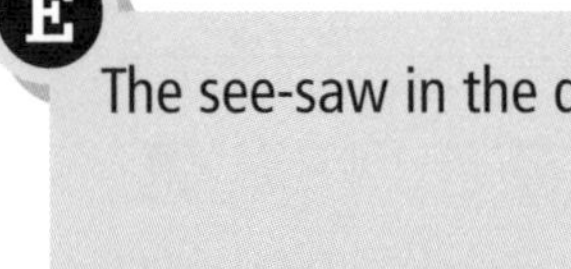

E

The see-saw in the diagram is balanced. What is the value of *x*?

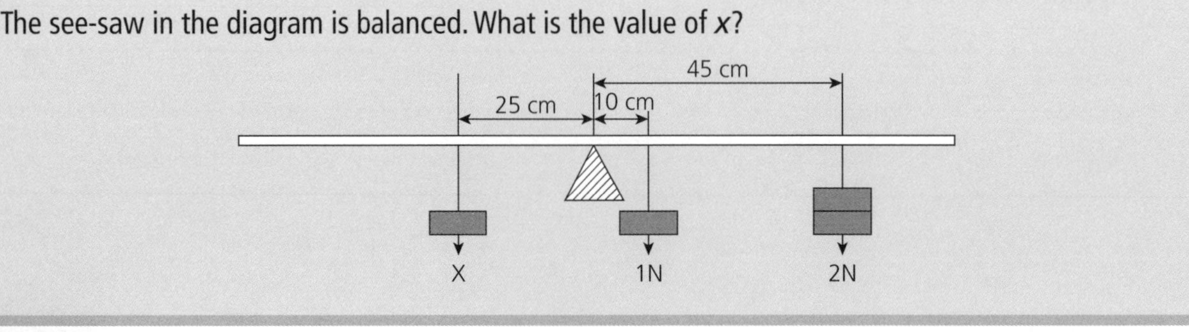

8.13 Planning

1 A student is playing with a mass that is connected to the end of a spring. He pulls the mass down and lets it go. It goes up and down. The time it takes for the mass to go up and down is called the period. Here is his plan.

I have decided to change the mass on the end of the spring.

I will make the mass bigger and measure the time.

I will put my results in a table like this.

Mass	Time

a What is wrong with the table?

....

b Make a list of improvements that he could make that would improve the data he collects.

....

....

....

c You have been given the following equipment:

- a box of springs of different length
- ten 100 g masses
- a stopwatch.

Plan an investigation to answer a different question from the one that the student was asking above. Write your plan in the box below.

E

a What can you do with your equipment *before* you start your investigation?

b Write down two reasons why doing that is a good idea.

8.14 Centre of mass and stability

1 A student does an experiment with a plastic bottle. He tips the bottle through different angles and measures the angle at which it topples over. Then he adds water to the bottle and measures the angle again.

a Here are the results of his experiment. Plot the results on the graph paper.

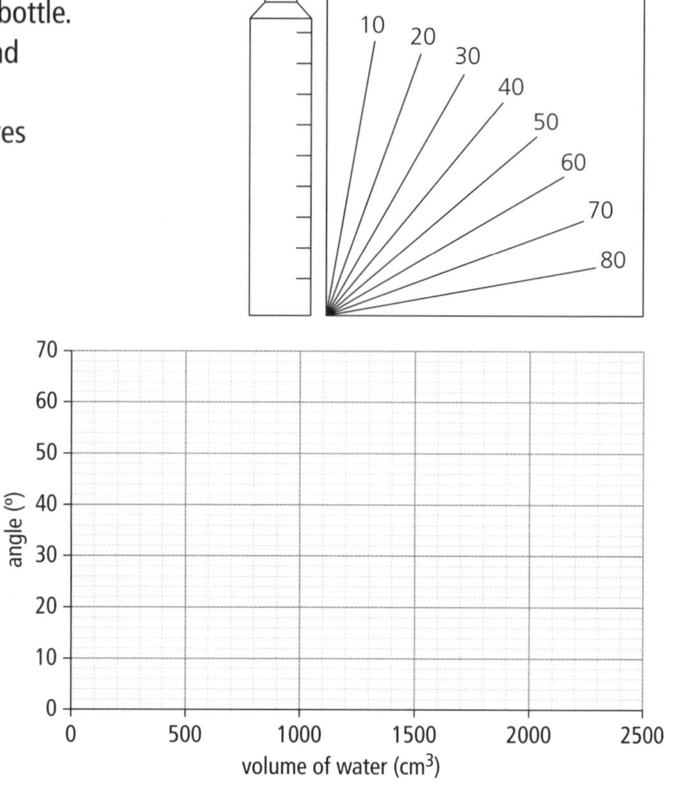

Volume of water (cm³)	Angle (°)
250	60
500	50
750	40
1000	30
1250	20
1500	15
1750	10
2000	10

b Write a conclusion for this experiment. Include these words in your conclusion: stable, centre of mass.

....

....

....

2 Boats come in many different designs. Many boats have large amounts of ballast, or heavy weights in the bottom.

Explain why.

....

....

3 This a diagram of a child's toy. The centre of mass is marked with a C.

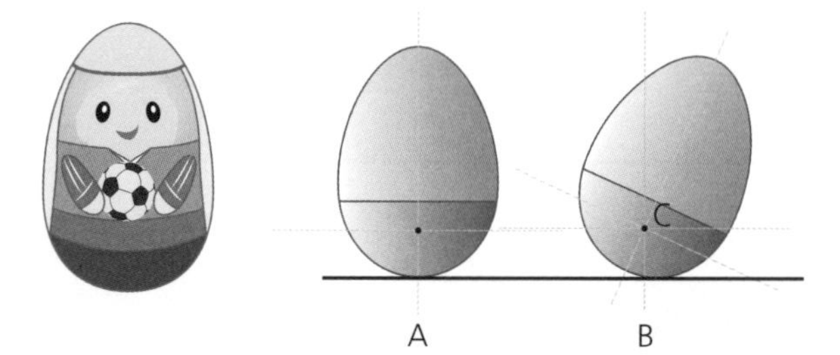

The manufacturers of the toy say that it will never topple over. They are correct. Explain why.

....

....

....

1 Use the words and phrases from the box to complete the sentences below.
Use each word once, more than once, or not at all.

electrons	conductors	iron	negative	insulators	positive
protons	neutrons	no	some	plastic	friction

a Static electricity is the imbalance between .... and .... charge.

It is caused by the movement of ....

b .... are materials that don't hold their electrons tightly and the electrons can easily move through them. An example is ....

c .... are materials that hold their electrons tightly and the electrons do not move easily. An example is ....

2 A student charges a rod and holds it over some small pieces of paper.

a Is the rod positively or negatively charged?

....

b How do you know?

....

c Draw a plus (+) or a minus (–) sign on the top piece of paper below the rod.

d Explain why you have chosen the sign that you have.

....

....

e Explain why the paper is attracted to the rod.

....

....

3 Here are some questions about using electrostatics.

a When you spray a car with paint the car is charged negatively.

i What is the charge on the paint? ....

ii What will happen when you spray the paint at the car?

....

iii What is one advantage of spraying a car in this way?

....

b You can clean smoke coming out of a chimney.
The smoke particles move through the grid and collect a negative charge. The smoke particles are then attracted to a collecting plate. What is the charge on the collecting plate?

....

1 Write down the definitions of these words.

a earthing ....

b spark ....

c current ....

d risk ....

2 Look at the diagram of the filling nozzle being used to fill up a car with petrol (gas or gasoline).

a Why can flowing petrol / gas become charged up?

....

b Sometimes a spark can jump from the nozzle to the car. Why would a spark be dangerous?

....

c If you hold the nozzle against the edge of the filling hole you are less likely to produce a spark. Why?

....

d Some petrol / gas nozzles have a metal wire connected from the nozzle to the ground along the fuel line. Explain how that would reduce the chance of producing a spark.

....

e Explain how your answers to parts **c** and **d** reduce the risk of damage to drivers or their cars.

....

3 a Describe how doing putting a lightning conductor on a building reduces the risk of damage.

....

....

b Describe how sitting inside your car in a thunderstorm reduces the risk to you.

....

....

E

Read the information in the box below.

> Lightning and sparks are examples of situations where there is an electric current flowing. The current does not flow through a metal but flows through the air. The air conducts. We talk about different things being insulators but there is only really one perfect insulator, which is a complete vacuum. It is very dangerous to be out in a thunderstorm because you could be struck by lightning, and the current would go through you to earth.

a Are there charged particles in the air? Explain your answer.

b Why is a vacuum an insulator?

c Which sentence above shows that you contain charged particles?

9.3 Digital sensors

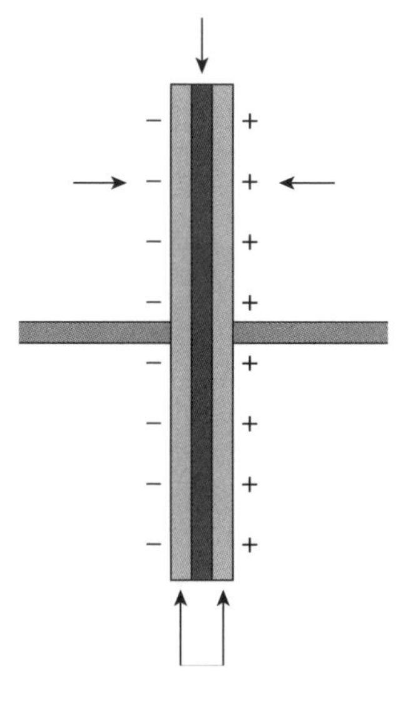

1 Here is a diagram of a capacitor.

a Label the diagram with the following labels:

plates   positive charge   negative charge   dielectric

b What has moved to produce the charge on the plates that you see in the diagram?

....

c If the dielectric becomes damaged it can 'break down' like the air does when lightning strikes. What would happen to the charge on the plates?

....

2 Circle the correct word or phrase in each **bold** pair in the paragraph below.

A touch screen on a phone is an example of a device that uses a **capacitor / CCD**. When you bring your finger close to the screen your finger acts as one of the **plates / terminals** and a sensing circuit works out the **position / temperature** of your finger. The camera on a phone is an example of a device that uses a **capacitor / CCD**. Light hits a grid of devices that produce **charge / current**. Each of the squares is a **capacitor / pixel**.

3 Here are some statements about how an image is produced and processed in a digital camera.

A Starting in one corner the charge on each square is moved off the CCD.

B Light enters the camera.

C That signal can be stored or sent to a computer or other device.

D The charges on each square are converted into a digital signal.

E When the light hits a square on the CCD charge is produced.

a They are in the wrong order. Write the letters for the statements in the correct order.

b Explain why there is a delay between taking one photograph with a digital camera and taking the next photograph.

....

....

E

A teacher is showing the class an experiment with electric fields. She connects up two plates to a power supply, and puts a candle between the plates.

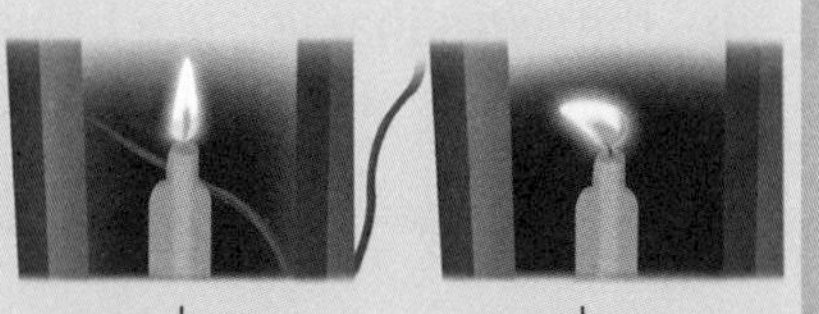

no charge on plates   charge on plates

a What is an electric field?

b How can you tell that the candle flame contains charged particles?

c The charge on the plate on the left is negative, and the charge on the plate on the right is positive. What is the charge on the particles in the flame?

d The teacher carefully moves the candle around between the plates. The candle looks the same. What can you say about the size of the electric field between the plates?

1 Draw lines to match each component to its picture and symbol.

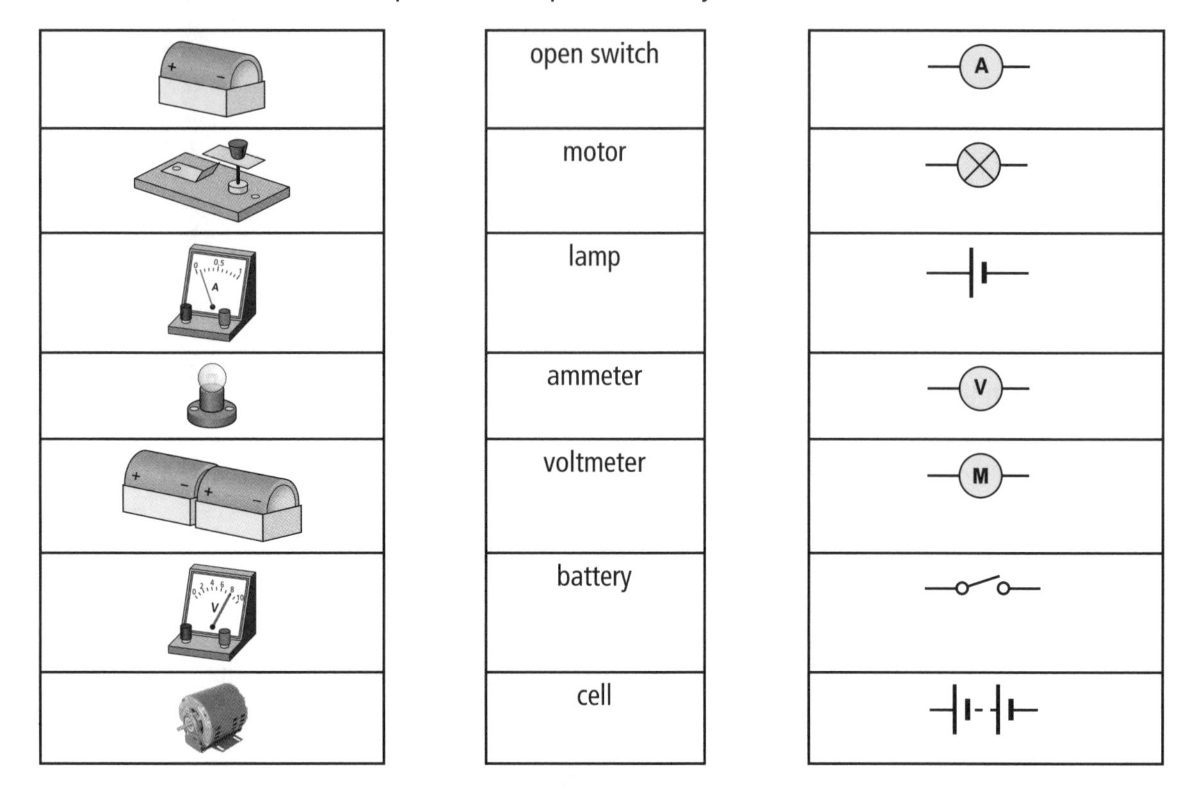

2 Here is a list of objects. Which will conduct electricity? Write each object in the correct columns in the table.

metal spoon wooden spoon plastic spoon piece of graphite (carbon) aluminium foil iron nail paper cup plastic bag

Conductors	Insulators

3 Explain why:

a The wires that you use to make circuits are made of metal. ....

b The wires that you use to make circuits are covered in plastic. ....

c The pins of a TV plug are made of metal. ....

d The outside of a plug is made of plastic. ....

E

A student makes a circuit to test whether a material conducts electricity using a cell, a lamp, and some wires.

a Draw a circuit diagram and explain how it works.

b Another student says that if the voltage is big enough even the air will conduct electricity. Is he correct? Explain your answer.

9.5 Current: what is it and how can we measure it?

1 Match the beginning of each sentence with the correct end of the sentence.
Connect the boxes together with lines.

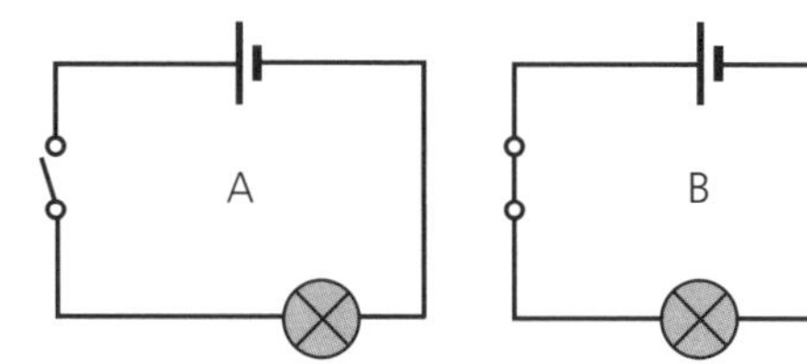

The current in a wire is …
Inside a metal wire …
There are 1000 milliamps …
You measure current …
The battery …

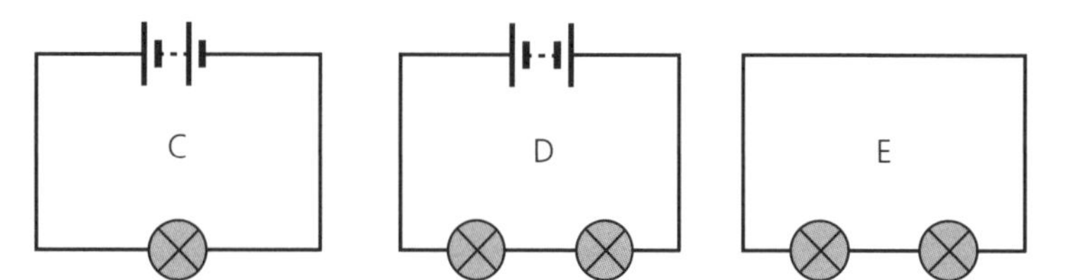

… in amperes, or amps.
… the charge flowing per second.
… provides the push to make the electrons in a wire move.
… in one amp.
… there are lots of electrons that move.

2 Look at the circuits below.

A B C D E

a In which circuits are the bulb or bulbs lit?

....

b For each circuit in which the bulb or bulbs are *not* lit, explain what you could do to make them light.

....

....

....

3 Which of these statements about current in a series circuit is true?
Write ***T*** next to the statements that are true. Write ***F*** next to the statements that are false.

a The current is bigger closer to the battery.

b If there are two bulbs in a circuit the first one will use up more current than the second one.

c The reading on an ammeter on one side of the battery will be the same as the reading on the ammeter on the other side of the battery.

d Each bulb will use up a little bit of current.

4 In this circuit the reading on ammeter 1 (A_1) is 2 A.

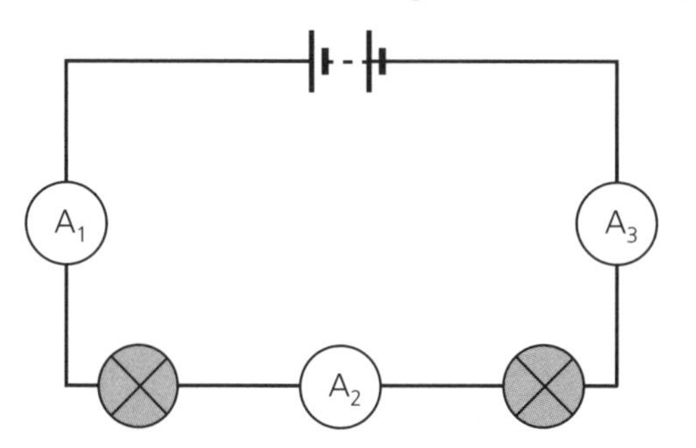

a What is the reading on ammeter 2 (A_2)? ....

b What is the reading on ammeter 3 (A_3)? ....

c What can you say about the current through the battery? ....

9.6 Parallel circuits

1 Here are some problems that students want to solve by building the correct circuit.
Draw the circuit diagram that they could use in the boxes below.
You need to decide which are series circuits and which are parallel circuits.

I want to make a torch that can be switched on or off and has one bright bulb.

I want to have a pair of lights in my bedroom that turn on together. I need them both to be bright and I don't want both going off if one bulb blows.

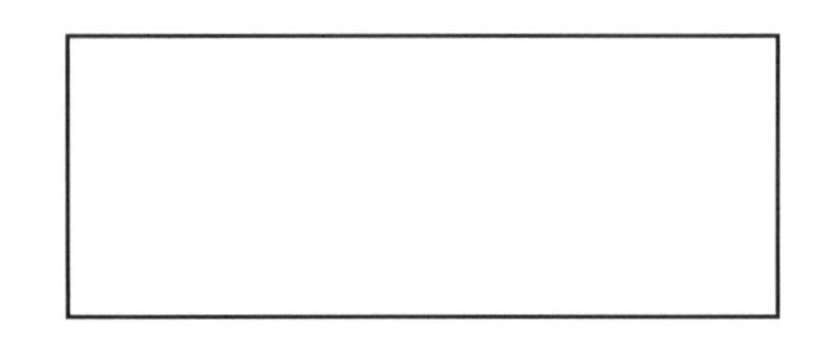

I am wiring two lights in my daughter's room. I'd like her to be able to turn the lights on and off individually but I want my own switch that turns them both off.

2 Here are some parallel circuits.

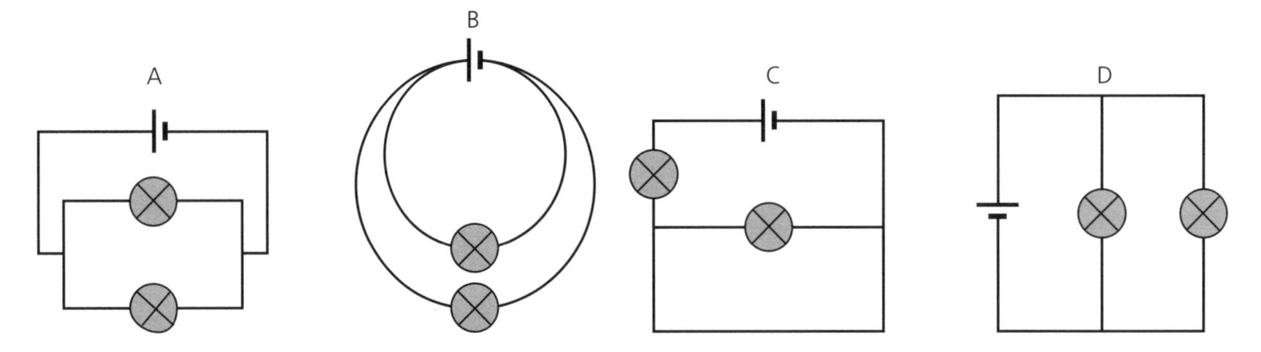

a Which circuit is the odd one out? ....

b Why? ....

3 A student has made some measurements of the current in three branches of a parallel circuit and the current near the battery. She investigated three different circuits A, B, and C. Her table is not complete.

Circuit	Current in branch 1 (A)	Current in branch 2 (A)	Current in branch 3 (A)	Current next to battery (A)
A	0.1	0.2		0.5
B		0.3	0.4	0.8
C	0.6	0.1	0.4	

a Complete the table.

b Explain why the current is different in the branches of each circuit.

....

....

1 A student wants to make a model for a series circuit using a bicycle. He starts to fill in a table comparing the bicycle model with a water model for a circuit with a bulb, ammeter, and battery. In the water model the pump pumps the water up and it falls back down.

a Complete the table.

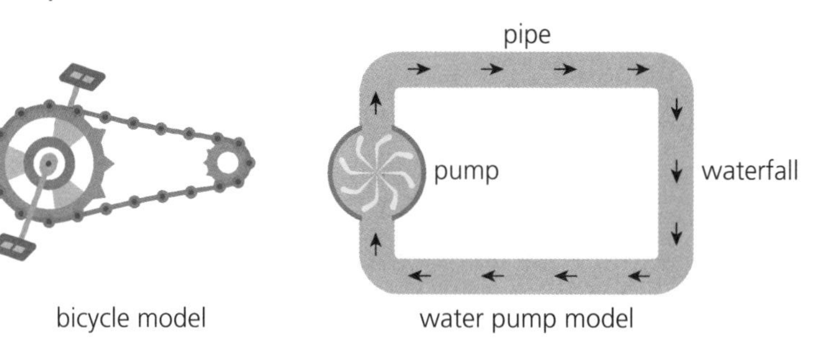

Circuit	Water model	Bicycle chain model
electrons		chain links
battery	pump	
conducting wire	pipes	
bulb	vertical pipes	
current		speed of chain motion

b How would you put a switch in the water model?

....

c How would you put a flat battery in the bicycle model?

....

2 A student describes how she models circuits using the rope model.
In each case draw the circuit diagram in the space provided.
Then write down what you would see happening if you connected the circuit.

Student 1 pulls the rope and student 2 and student 3 hold the rope. Student 2 holds more firmly than student 3.	Student 1 pulls the rope in one direction, and student 2 pulls the rope in the opposite direction. Student 3 holds the rope.
Circuit diagram:	Circuit diagram:

E

Here is a circuit diagram for a parallel circuit.

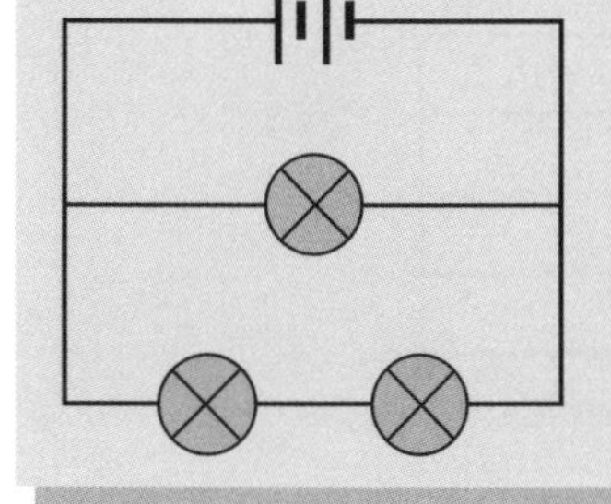

Explain how you would model this circuit with:

a students carrying sweets.

b the rope model.

1 Circle the correct word or phrase in each **bold** pair in the paragraph below.

In a series circuit increasing the number of bulbs will make all the bulbs **brighter / dimmer / the same brightness**. In a parallel circuit if you add another bulb the other bulb or bulbs will be **brighter / dimmer / the same brightness**. If you add another cell in a series or parallel circuit the bulbs will be **brighter / dimmer / the same brightness**. If you add more bulbs in a series circuit you are making the resistance **bigger / smaller / the same**, and this will make the current **bigger / smaller / the same**. In a parallel circuit the current in the branches **adds up / cancels out** to the current through the battery.

2 Look at the circuits below. In which of the circuits are the bulbs 'normal brightness'? For circuits A–D give the circuit and the number of the bulb (e.g. A1). For circuits E and F just give the letter (e.g. F).

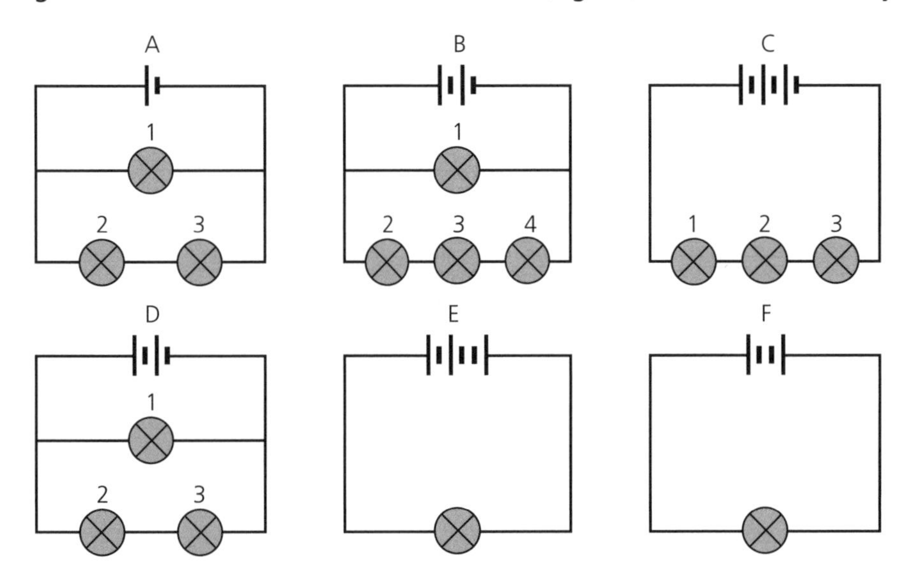

....

E A student connects up a circuit with a bulb of normal brightness. This is circuit A.

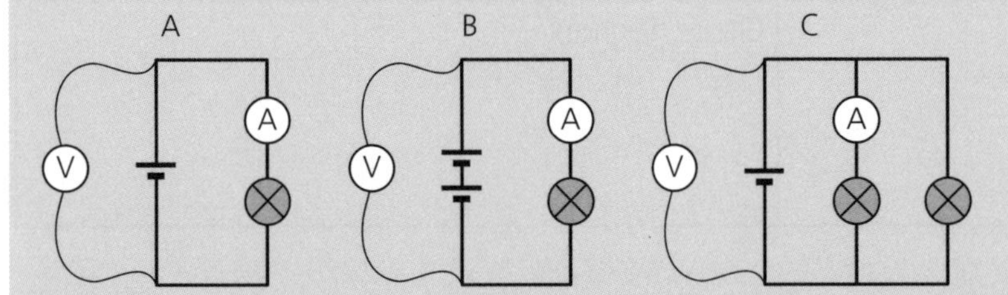

Copy and complete the following table to show how the brightness of the bulbs and the readings on the meters, is different in circuits B and C compared with circuit A.

	In B compared with A	**In C compared with A**
the brightness of the bulb or bulbs		
the reading on the voltmeter		
the reading on the ammeter		

1 Use the words and phrases from the box to complete the sentences below.
Use each word once, more than once, or not at all.

current	voltage	energy	volts	push	pull	charges	resistance	voltmeter	ammeter

The .... of a cell tells you the size of the .... that the cell will give to the .... that flow in the wires. The .... flowing produce a .... If the voltage is bigger, more .... will be transferred to the components by the .... Voltage is measured in .... using a ....

2 Here are some ways of connecting a voltmeter in a circuit.
The battery has two cells, and each cell has a voltage of 3 V.

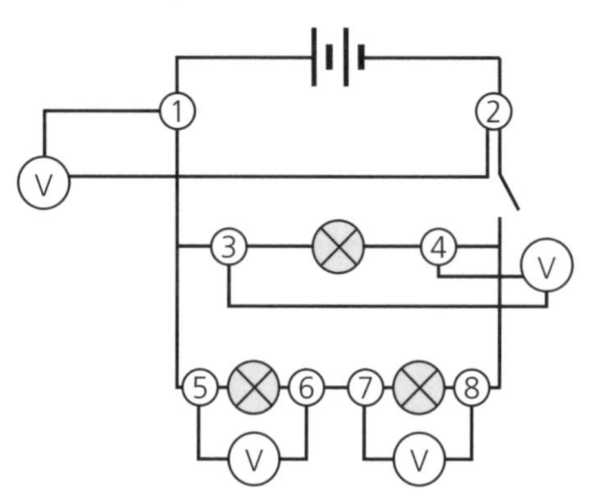

a What is the reading on the voltmeter connected to positions 1 and 2?

....

b What is the reading on the voltmeter connected to positions 3 and 4 when the switch is open?

....

c What is the reading on the voltmeter connected to positions 5 and 6 when the student closes the switch?

....

3 Write ***T*** next to the statements that are true. Write ***F*** next to the statements that are false.

a A voltmeter is always connected in series with a component.

b A voltmeter is used to measure voltage.

c You measure the voltage through a component and the current across a component.

d The voltages across two components in series will add up to the voltage across the battery.

e Two cells of 1.5 V each would always add up to 3 V.

E

Look at the circuit below.

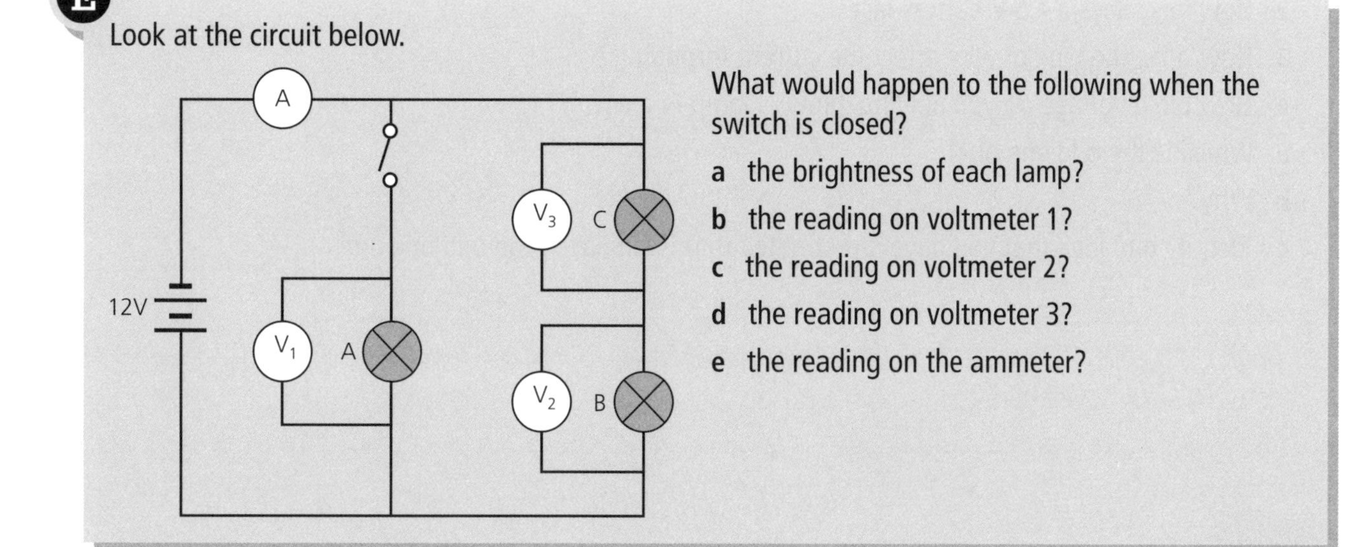

What would happen to the following when the switch is closed?

a the brightness of each lamp?

b the reading on voltmeter 1?

c the reading on voltmeter 2?

d the reading on voltmeter 3?

e the reading on the ammeter?

9.10 Selecting ideas to test circuits

1 Circle the ideas below that you can test by collecting data.

a How does the current change if I change the number of bulbs in a series circuit?

b Which battery is better?

c Does the time it takes for a light bulb to blow depend on how many times you turn it on and off?

2 A student decides to investigate the effect of changing the current in a coil of wire.
Wires get hot when a current flows through them.
He decides to put a coil of wire in a beaker of water and measure how hot the water gets.

Complete the table below to show how he can do this investigation.

Independent variable	
Dependent variable	
Variables that he needs to control	
Preliminary work he needs to do	
Equipment that he will need	
Diagram showing how he will set up the equipment	
Hazards and how to control risk	
Table of results that he will need to draw	

Two students are discussing different ideas that they could investigate. Here is a list of those questions.

1 Which is the most powerful light bulb?

2 How long does a 1.5 V battery last?

3 How does the type of wire affect the current through it?

4 How big a voltage do you need to 'blow' a lamp in a circuit?

a Which is the odd one out?

b Why?

c Identify one idea that you have investigated that is similar to the odd one out.

9.11 Energy and power

1 Write ***T*** next to the statements that are true. Write ***F*** next to the statements that are false.
Then write corrected versions of the statement or statements that are false.

a Lamp A that emits 500 J of light energy in 10 seconds is more powerful than lamp B that emits 60 J in 1 second.

b A motor that produces 750 J of kinetic energy in one second is less powerful than a motor that produces 1200 J in 2 seconds.

c A buzzer that emits 250 J of sound energy in 5 seconds has the same power as a buzzer that emits 200 J in 4 seconds.

d A torch bulb that emits 50 J of light energy in 10 seconds is less powerful than a torch bulb that emits 75 J in 15 seconds.

e There are fewer watts in a kilowatt than in a watt.

Corrected versions of false statements:

....

....

2 Calculate the power of each device and complete the table.

Device	Energy (J)	Time	Power
shower	60 000	1 minute	
refrigerator	1 800 000	1 hour	
low energy lamp	252 000	7 hours	
hairdryer	576 000	12 minutes	

3 **a** How much energy in joules does a 10 kW shower use per second?

....

....

b How much energy in joules would a person having a shower use in 10 minutes?

....

....

c How long would it take the same shower to use 20 kJ?

....

....

4 A girl uses a hairdryer to dry her hair. The power of the hairdryer is 800 W.

a What is the power in kW?

....

b She uses the hairdryer for 30 minutes per week. How many hours is 30 minutes? Calculate the energy in kWh.

....

c It costs 10 rupees per kWh. How much does it cost to use the hairdryer?

....

10.1 Hot and cold

1 Use the words and phrases from the box to complete the sentences below.
Use each word once, more than once, or not at all.

energy	longer	shorter	temperature	mass	colour

The time it takes to heat some water depends on the .... of water you are heating, and the .... that you want to the water to reach. It takes a .... time to heat a greater .... of water because it needs more .... It would also need more .... to heat the same .... of water to a higher ....

2 Explain why:

a It takes longer to heat a cinema to the same temperature as a house.

....

b It takes longer to boil a pan of water than to heat a pan of water to 50 °C.

....

c Rewrite this sentence so that it is correct: 'The energy stored in the Sun is 6 million degrees C.'

....

3 Complete the table by ticking the one or both columns for each statement.

	Thermal energy	Temperature
Measured in joules.		
Measured with a thermometer.		
Does not depend on how much material there is.		
Measured in degrees Celsius.		
Increases if you heat something for longer.		

4 It takes 4200 J to raise the temperature of 1 kg of water by 1 °C.

a How much energy in kJ would it take to raise the temperature of 1 kg of water by 2 °C?

....

b How much energy in kJ would it take to raise the temperature of 3 kg of water by 1 °C?

....

E

It takes 4.2 kJ to raise the temperature of 1 kg of water by 1 °C, but it only takes 2.1 kJ to raise the temperature of 1 kg of cooking oil by 1 °C. Which of the statements below is true?

a It takes more energy to heat 1 kg of oil from 20 °C to 30 °C than it does the same mass of water from 20 °C to 30 °C.

b 1 kg of oil would reach a higher temperature than 1 kg of water if heated for the same length of time. (Assume the starting temperature is the same.)

c It takes roughly twice as much energy to raise the temperature of 1 kg of water by 10 °C as it does to raise the temperature of 1 kg of oil by 10 °C.

10.2 Energy transfer: conduction

1 Write ***T*** next to the statements that are true. Write ***F*** next to the statements that are false. Then write corrected versions of the statements that are false.

a Liquids are poor conductors.

b Things that feel warm conduct thermal energy away from our hands.

c Wood and plastic are insulators.

d The particles in a metal that is hot are vibrating less than the particles in a metal that is cold.

Corrected versions of false statements:

....

....

2 A student is investigating what makes a good insulator. She wraps different materials around a can of hot water and measures the temperature drop of the water over 10 minutes. Here are her results.

Material	Temperature drop (°C)
paper	20
cotton	14
foam	6
wool	18

a Which material is the best insulator?

....

b Which material is the worst insulator?

....

c The student cuts a piece of the best insulating material. She notices that there are lots of pockets of air in it. Explain why that will help the material to insulate the cup.

....

3 A teacher is showing her students how metals conduct. She dips the ends of three metal rods in melted wax and then sticks a drawing pin to the end of each rod. She heats the other ends of the rods with a Bunsen burner.

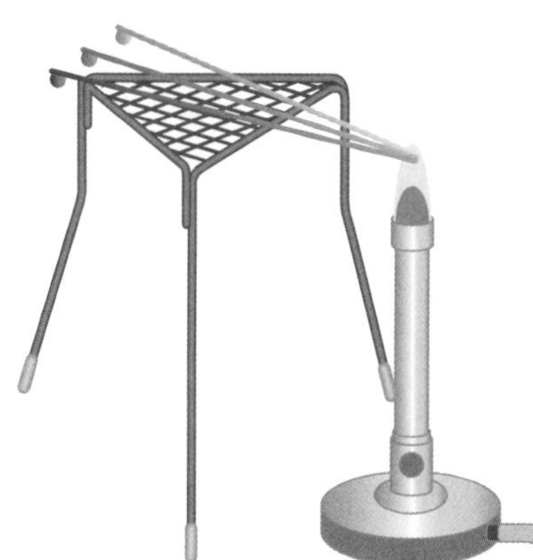

Metal	Time for drawing pin to drop off (s)
copper	60
iron	380
aluminium	200

Are these statements true or false?

a The drawing pin drops off the rod that is the best conductor first.

b If the rod was made of an insulator the drawing pin would not drop off.

c Look at the times that it took for the drawing pins to drop off.
Put them in order of best conductor to worst conductor.

....

1 Use the words and phrases from the box to complete the sentences below.
Use each word once, more than once, or not at all.

upwards	colder	decreases	expands	convection current
	increases	denser	warmer	less dense

When air is heated it .... This means that its density .... It moves .... and ...., ... air moves in to take its place. This movement of air is called a ....

2 A student is doing an experiment about convection. She puts a purple crystal in the bottom of a beaker of water and heats it from below.

a On the diagram draw what she would see after a few moments.

b What is produced in the water when it is heated?

....

c She takes another beaker and fills it with water.
This time she puts an ice cube made with purple water on top.
On the diagram draw what she would see after a few moments.

ice cube

3 A teacher is demonstrating convection in air.
The diagrams show an experiment that demonstrates convection in air.
The smouldering rag produces smoke.

a Add to diagram 1 what you would see soon after the experiment started.

b Add to diagram 2 what you would see when the experiment has been running for several minutes.

smouldering rag or taper

smouldering rag or taper

c Explain why you have drawn what you have on the drawings.

....

....

....

E

Explain why:

a In an experiment to compare how well different materials insulate a beaker of water, you must put a lid on the beaker.

b A bottle of water will cool quicker if you put it in the bottom of a refrigerator rather than the top.

c Food cooks quicker at the top of an oven rather than at the bottom.

1 Read the information in the box then answer the questions below.

> You can help to locate survivors of an earthquake using video cameras and thermal imaging cameras. A video camera on the end of a flexible pole is squeezed through gaps in the rubble to help locate survivors. A thermal imaging camera can be used to find people as their body heat warms the rubble around them.

a What is the difference between a video camera and a thermal imaging camera?

....

b Explain why it would be more difficult to find someone using a thermal imaging camera if parts of the building were burning.

....

c Would a thermal imaging camera 'see' someone trapped behind a metal door? Explain your answer.

....

d Describe one advantage of using a thermal imaging camera compared with a video camera.

....

2 The Earth's sister planet is Venus. Venus is about the same size as the Earth but it has a very different atmosphere. The planet Mercury has no atmosphere.

	Earth	**Venus**	**Mercury**
atmosphere	78% nitrogen 21% oxygen 1% other gases including carbon dioxide	96% carbon dioxide 3% nitrogen 1% other gases	none
density of atmosphere compared with Earth	1	56	0
temperature	20 °C (–80 °C to +50 °C)	462 °C	–183 °C to +467 °C

a Use the information in the table to explain why the average temperature on Venus is so much higher than on Earth.

....

....

b Use the information in the table to explain why the temperature on Mercury varies so much more than the temperature on Earth.

....

....

E

a Copy and complete the table of the waves of the electromagnetic spectrum.

radio			visible			

b Gamma, X rays and ultraviolet are more dangerous to the human body than radio, microwaves and infrared. Suggest why.

1 Write ***T*** next to the statements that are true. Write ***F*** next to the statements that are false. Then write corrected versions of the statements that are false.

a The average speed of molecules in a liquid is the fastest speed.

b When a liquid evaporates the average speed of the molecules in the liquid decreases.

c If the temperature of a liquid becomes lower, the average speed of the molecules in it will be bigger.

d All liquids evaporate at the same rate.

Corrected versions of false statements:

....

....

2 A nurse in a hospital is giving a child an injection. She wipes some ethanol (alcohol) on the child's arm before she gives the injection.

a Why does the child's arm feel cold when she puts ethanol on it?

....

b Water on your arm doesn't feel as cold as ethanol on your arm. Explain why.

....

3 A teacher demonstrates what happens when a liquid evaporates. He sets up this experiment. Blowing air through ether (a special chemical) makes it evaporate quickly.

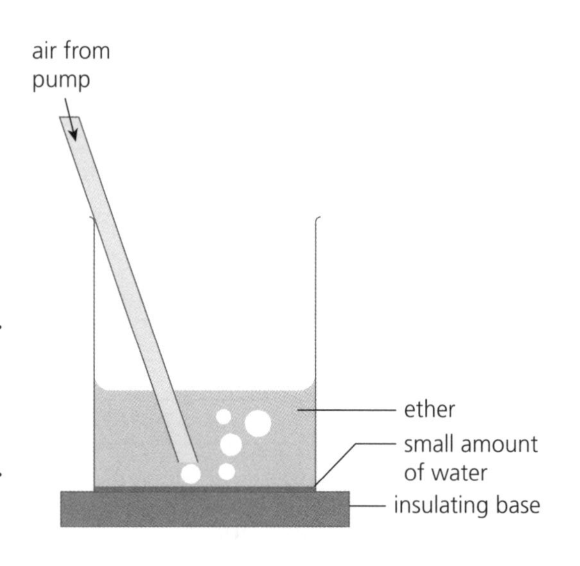

a After a while the water under the beaker turns to ice. Why?

....

b How does this experiment show what happens in a refrigerator?

....

E

Here is a graph that shows the number of molecules with different speeds in a liquid at two different temperatures.

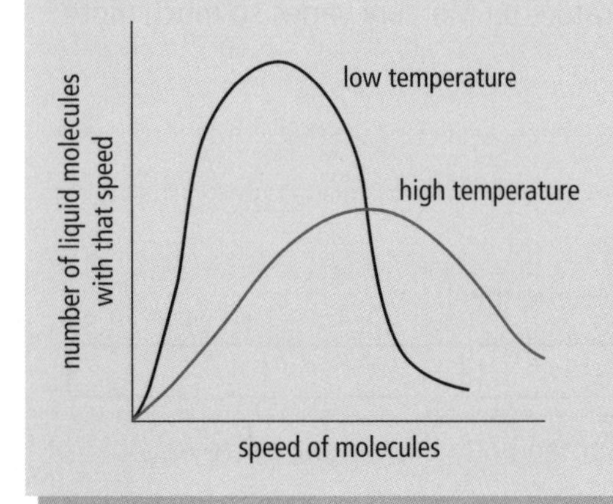

a Explain what the graph shows about the speeds of molecules in a liquid at a lower temperature compared with the speed of molecules at in a liquid a higher temperature.

b Is the average speed of the molecules higher or lower at a higher temperature?

c Explain why warm water evaporates faster than cold water.

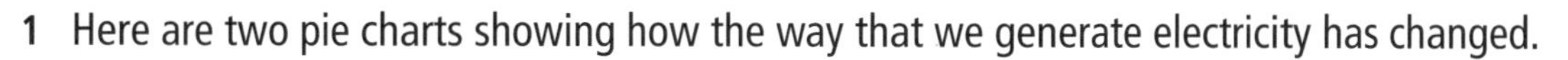

1 Here are two pie charts showing how the way that we generate electricity has changed.

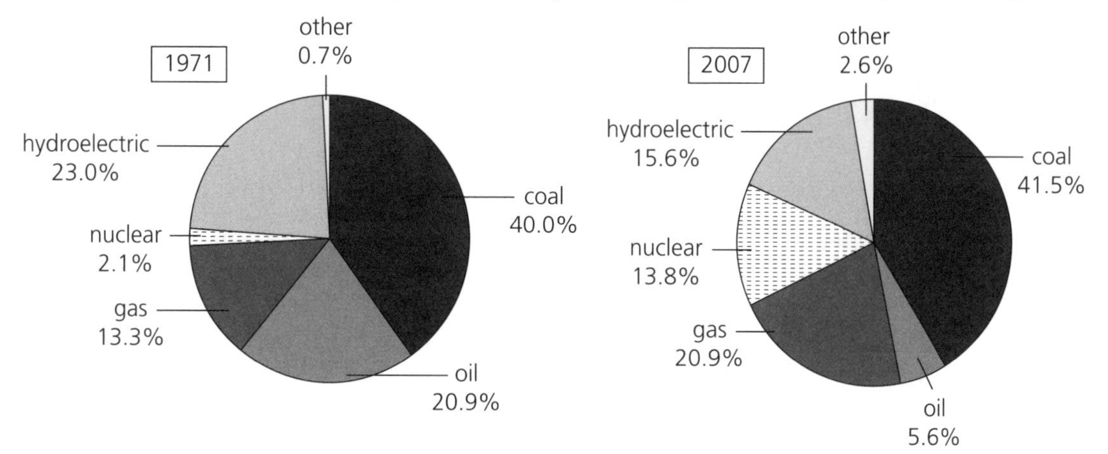

a Which of the fuels in the pie charts are renewable (ignoring 'other')?

....

b Which of the fuels in the pie charts are non-renewable (ignoring 'other')?

....

c Look at the percentages on each pie chart and complete the table below. (*Hint* – the change in use will be negative if use has decreased, and positive if it has increased.)

Fuel	Percentage use in 1971	Percentage use in 2007	Percentage change in use
coal			
oil			
gas			
nuclear			
hydro			

d List the fuels in order from the biggest drop to the biggest increase (ignoring 'other') in use.

....

e A student thinks that the use of non-renewables is decreasing and the use of renewables is increasing. Is he correct? Explain your answer.

....

....

f If the change continues, which fuel will we not be using at all 20 years from now?

....

g Suggest one way of generating electricity that could be in the category 'other'.

....

E

Look at the pie charts above.

a How many years are there between the two dates?

b Calculate the percentage change per year for each fuel by dividing the percentage change by the number of years. Round up your answer to 2 decimal places.

10.7 Fossil fuels

1 Use the words and phrases from the box to complete the sentences below.
Use each word once, more than once, or not at all.

animals	thousands	mud	trees	compressed	millions	sand	rock	stretched

Coal is made from .... which grew in swamps .... of years ago.

When they died they fell into the swamp and were covered in .... More layers of

.... formed and .... them and heated them. Over .... of

years the .... turned to rock and the .... turned to coal.

2 Complete the table by placing a tick (✓) in the correct column.

	True for the coal	True for oil	True for both
Took a very long time to form.			
Made from trees.			
Made from sea creatures.			
Can be found underground.			
Formed as a result of heat and pressure.			

3 Here is a list of statements about how electricity is generated in a fossil fuel power station.
They are in the wrong order.

A The turbine drives a generator.

B Water is heated to produce steam.

C A fossil fuel is burned.

D The steam drives a turbine.

E Electricity is generated.

a Write down the letters so that the statements are in the correct order.

b Here is a Sankey diagram of fossil fuel power station. Complete the missing labels on the diagram.

c How efficient is the power station? ....

E

a Make a table that shows the pros and cons of using fossil fuels to generate electricity.
There should be at least three statements in each column.

b Why is it not possible to say exactly when fossil fuels will run out?

c Explain why it is not possible to make more fossil fuels.

1 Use the words and phrases from the box to complete the sentences below.
Use each word once, more than once, or not at all.

bigger	coil	current	direction	electromagnet	induced	magnet
magnetic	more	poles	speed	strength	stronger	

a You can generate electricity by moving a .... of wire in a .... field. The size of the voltage depends on the .... of the movement and the .... of the magnet.

b In a model generator a .... rotates between the .... of a magnet.
If a .... with .... turns is used the voltage is greater.
If the .... spins .... the voltage is greater.

2 A student moves a coil towards a magnet and sees an induced voltage on the voltmeter.
What would she notice about the induced voltage if:

a She moved the magnet towards the coil at a slower speed?

....

b She moved the magnet towards the coil at the same speed?

....

c She turned the magnet round and moved it towards the coil?

....

3 Here is a diagram of a simple model generator.

a Label the coil and magnets.

b Show how you would connect a voltmeter to measure the induced voltage.

c Which of these statements about the generator are true?
Write ***T*** next to the statements that are true. Write ***F*** next to the statements that are false.

i In a real generator the coils don't move but the magnet does.

ii A real generator contains a permanent magnet.

iii A real generator is much smaller than the model generator.

E

A teacher wants to demonstrate that the Earth has a magnetic field by generating electricity. She takes some students outside and they all line up. She gives them a *very* long piece of wire to hold and connects the end of the wire to a voltmeter. The students move their arms up and down and the needle moves.

a If the needle moves left when they move their arms downwards, which way does it move when they move their arms up?

b Why does the teacher need to use a very long piece of wire?

c Is there anything that the teacher could tell the students to do to make the reading on the voltmeter bigger?

1 Use the words and phrases from the box to complete the sentences below.
Use each word once, more than once, or not at all.

battery brighter closer copper dark dust electrical greenhouse light renewable Sun sunshine water

a The .... is the main source of .... for the Earth. A solar cell transfers .... energy into .... energy. The output of a solar cell increases when it is .... to the light source and when the light source is .... Solar energy is a .... energy source.

b Solar panels use the Sun's .... to heat .... flowing through copper pipes.

c Solar cells and solar panels do not produce .... gases that can cause climate change.

d The disadvantage is that they do not produce energy when it is ....

2 Kenya has lots of geothermal power stations.
Here is a diagram of a geothermal power station.

a Label the three arrows. Use these words or phrases: hot water, cold water, generator.

b A student thinks geothermal power stations do not contribute to climate change. Is that true?
What would you say to her?

....

....

c Some places are more suitable for geothermal power stations than others. (*Hint:* Think about what you have learned about the structure of the Earth's crust.)
Where is the best place to put one? Explain your answer.

....

....

3 Here are some advantages and disadvantages of renewable energy sources for generating electricity. Which apply to solar, which apply to geothermal, and which apply to both?
Tick (✓) the correct columns.

Advantage or disadvantage	Solar	Geothermal	Both
It is unreliable.			
It doesn't produce much carbon dioxide when running.			
It is expensive to produce/build.			
It won't run out.			

1 Match the beginning of each sentence with the correct end of the sentence. Connect the boxes together with lines.

Hydroelectricity …	… when they are being manufactured.
When water falls through turbines in a dam …	… can destroy habitats when valleys are flooded.
A tidal barrage contains turbines and generators …	… electricity is generated by wave power.
Wind turbines produce carbon dioxide …	… electricity (called hydroelectricity) is generated.
When water moves into a chamber on a shoreline …	… to generate electricity when the tide goes in or out.

2 A wind generator produces electricity when the wind is blowing. The table shows the electrical power generated by the wind for different wind speeds.

Wind speed (km/h)	0	2	5	8	11	13	15	20
Power generated (watts)	0	0	150	600	1000	1120	1170	1170

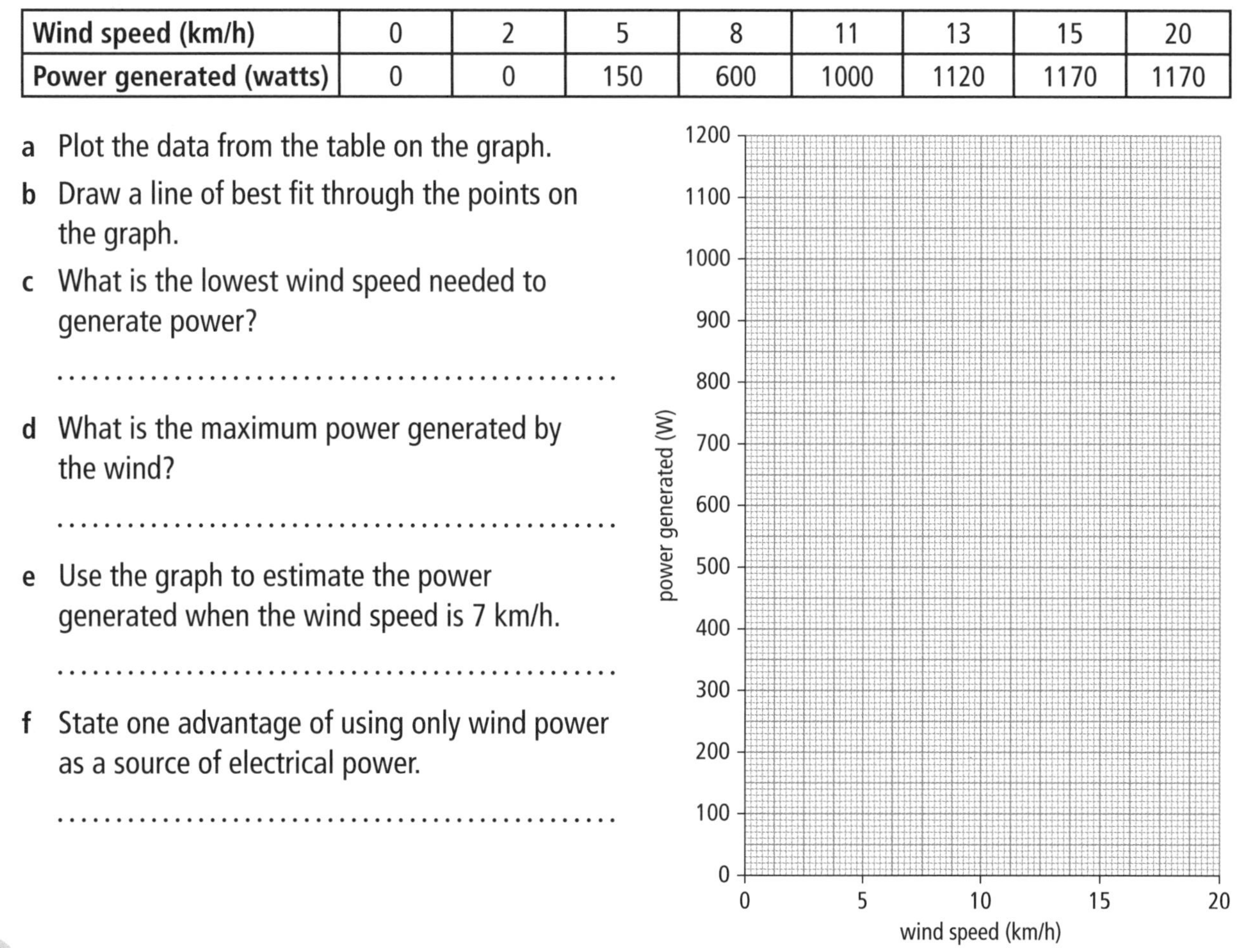

a Plot the data from the table on the graph.

b Draw a line of best fit through the points on the graph.

c What is the lowest wind speed needed to generate power?

....

d What is the maximum power generated by the wind?

....

e Use the graph to estimate the power generated when the wind speed is 7 km/h.

....

f State one advantage of using only wind power as a source of electrical power.

....

E

Here are some data about wind, wave, and tidal power.

	Wind	**Wave**	**Tidal**	**Coal**
power output of largest power station (MW)	1020	2.25	254	5780
typical power station (MW)	500	0.3	20	1000

a Why is it difficult to generate electricity from waves?

b Which of the renewable sources has the largest average output?

c Give two disadvantages of using this source to generate electricity.

d Why is it unlikely that there will be lots of tidal power stations in the future?

1 A farmer is thinking of installing some alternatives to generate electricity.
Calculate the payback time for each one.

Type	Cost to install (rupees)	Saving per year (rupees)	Payback time (years)
solar water heating	450	75	
solar cell (electricity)	35 000	2500	
wind turbine	50 000	5000	

2 Read the information in the box below about petrol (gas or gasoline), hybrid, and hydrogen fuel cell cars.

The cost of running cars that use petrol or gasoline depends on the price of crude oil. This affects the price at the pumps. Petrol / gas engines emit large amounts of carbon dioxide. This is a greenhouse gas and contributes to climate change.

Hybrid cars use a mixture of petrol / gas and electric batteries. Batteries cause very little pollution when used in a car. The batteries need to be charged from the electricity supply. If the electricity is generated in a fossil-fuel power station carbon dioxide is emitted at the power station.
If a renewable energy source is used much less carbon dioxide is emitted.

Electric cars run solely on rechargeable batteries. They need an electricity supply and the battery runs down quite quickly. Electric cars cannot travel very fast.

Some cars run on hydrogen. They emit only water. Most of the hydrogen fuel is made by splitting water into hydrogen and oxygen using an electric current. This means that cars that use hydrogen may still contribute to climate change. Hydrogen cars can have a top speed as high as a petrol / gas car but need frequent refuelling, and there are very, very few refuelling stations.

Here are some comments that people have made about the different types of car. Are they correct?
Write a response in the boxes below each statement.

Petrol / gas cars are the best. The alternatives can't be refuelled or go so fast.	Hybrid cars don't contribute to climate change.	Electric cars aren't as fast as petrol / gas cars, but they are easy to refuel.

E

A student makes some notes about solar cells using information from a website.

Total energy needed to make solar cells is 580 kWh. They produce about 160 kWh per year and last about 20 years. They produce about 230 kg of CO_2 during its lifetime. A wind turbine produces about 8 million kWh per year and 20 000 kg of CO_2 in its lifetime. It takes 4 000 000 kWh to make one wind turbine.

a Why is the CO_2 produced important?

b Compare and contrast wind power and solar cells using this data.

Practice questions

1 Look at the following list of resources.

Sun	oil	biomass	coal

a Name one fuel from the list.

.... [1]

b Which of the resources in the list are renewable?

.... [2]

c A wind turbine uses energy from the wind to generate electricity.
What do we call this energy? Circle the correct answer.

thermal **kinetic** **elastic** **chemical** [1]

d What is the name of the energy found in coal? Circle the correct answer.

thermal **kinetic** **elastic** **chemical** [1]

2 Navin and Mohan were carrying out a balancing experiment. They balanced a flat ruler on top of a pivot and then put two 1 N weights on either end.

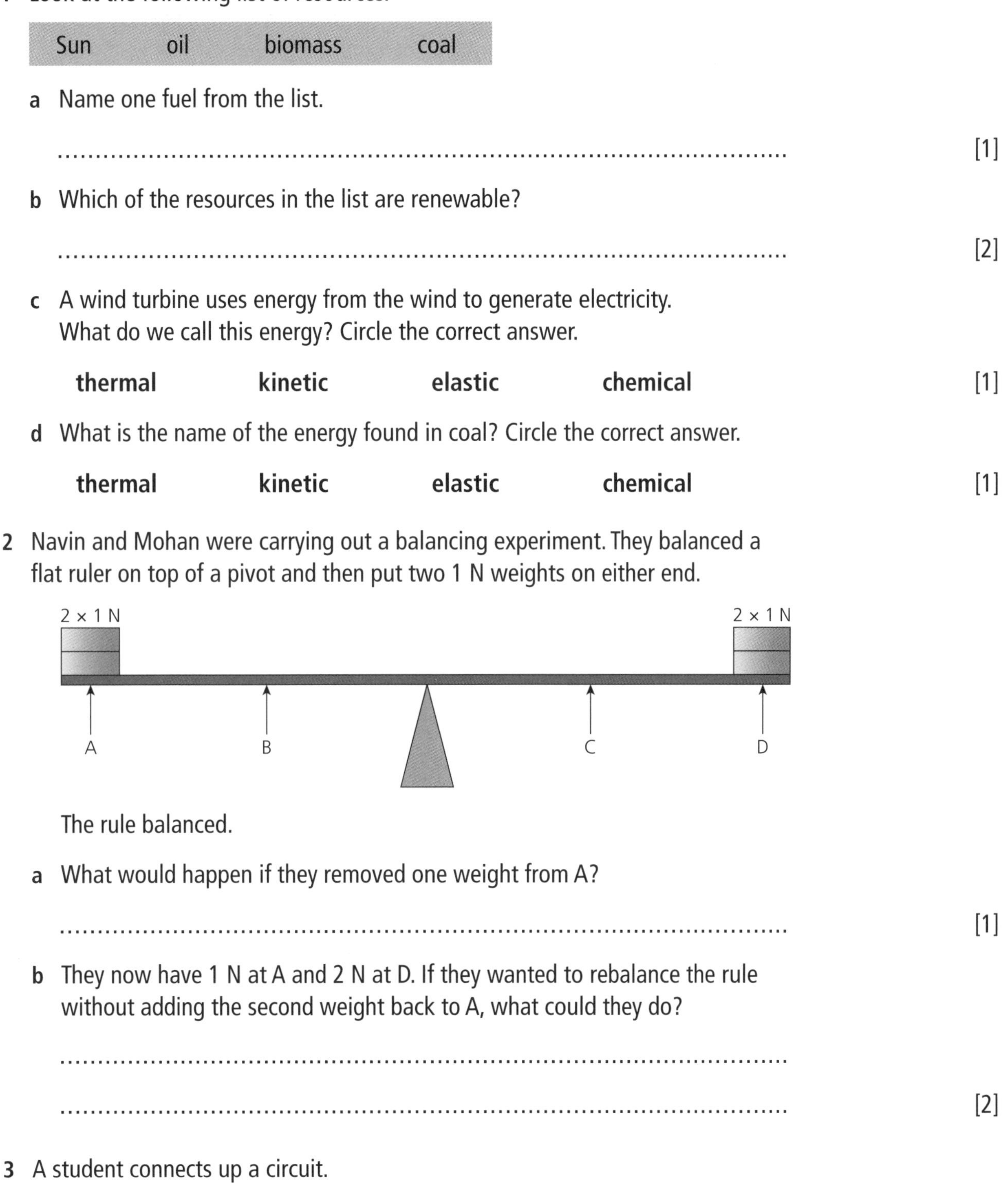

The rule balanced.

a What would happen if they removed one weight from A?

.... [1]

b They now have 1 N at A and 2 N at D. If they wanted to rebalance the rule without adding the second weight back to A, what could they do?

....

.... [2]

3 A student connects up a circuit.

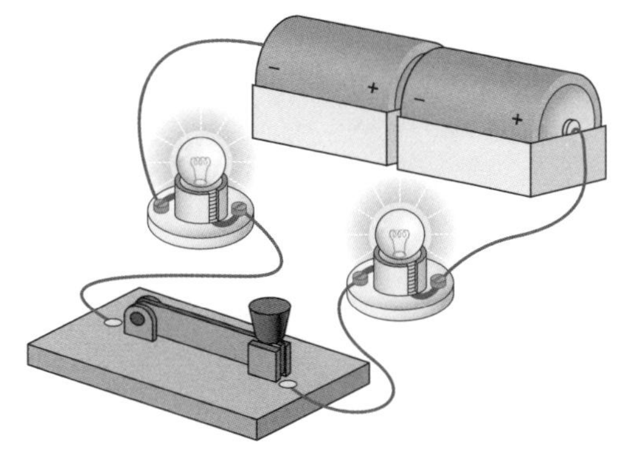

a Which of the three circuit diagrams below represents this circuit?
Circle the correct circuit diagram.

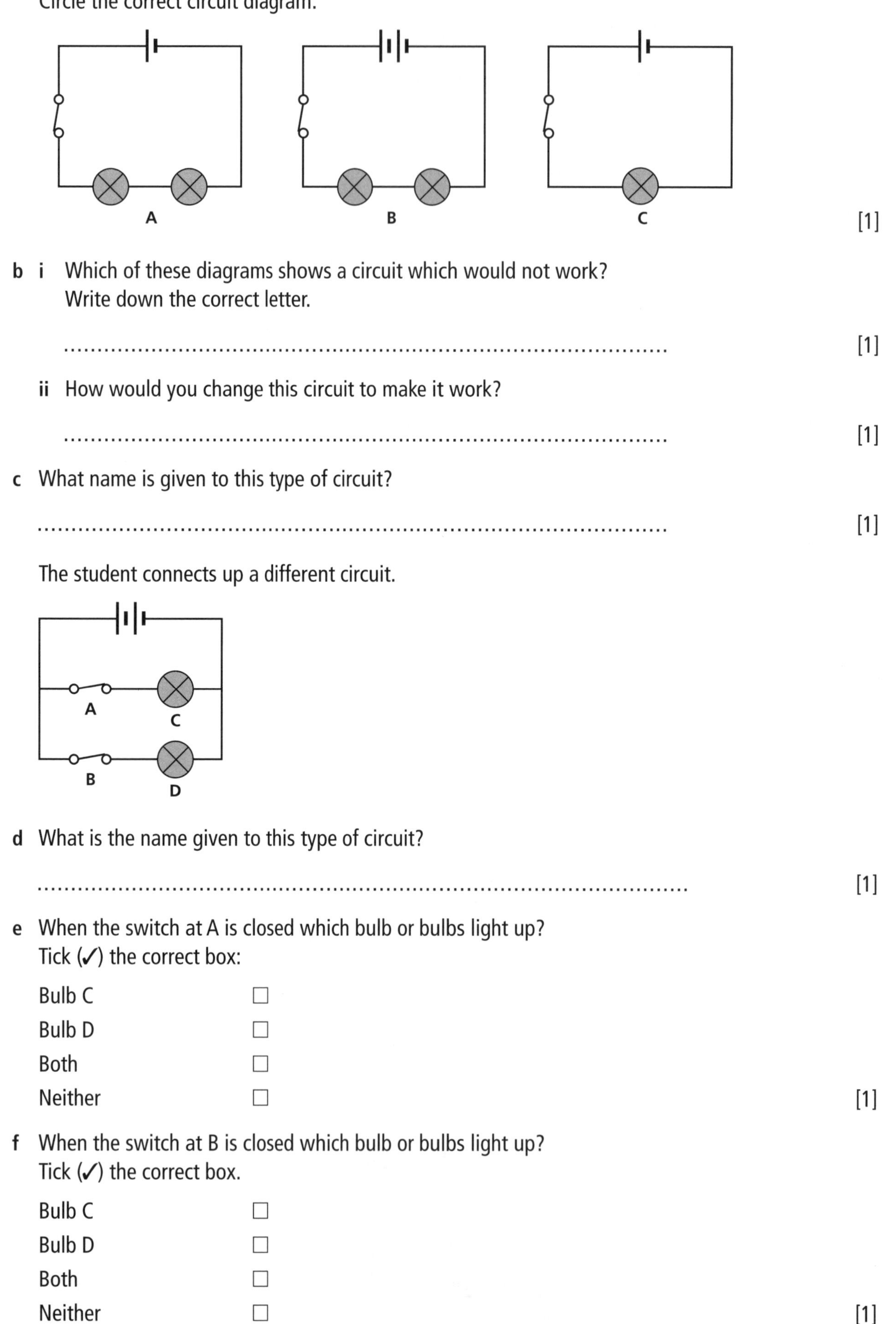

[1]

b i Which of these diagrams shows a circuit which would not work?
Write down the correct letter.

.... [1]

ii How would you change this circuit to make it work?

.... [1]

c What name is given to this type of circuit?

.... [1]

The student connects up a different circuit.

A C
B D

d What is the name given to this type of circuit?

.... [1]

e When the switch at A is closed which bulb or bulbs light up?
Tick (✓) the correct box:

Bulb C ☐
Bulb D ☐
Both ☐
Neither ☐ [1]

f When the switch at B is closed which bulb or bulbs light up?
Tick (✓) the correct box.

Bulb C ☐
Bulb D ☐
Both ☐
Neither ☐ [1]

4 A kettle transfers electrical energy to thermal energy.

This transfer can be shown by the energy transfer diagram below

electrical energy	kettle →	thermal energy

a A train also transfers energy from one place to another. Complete the diagram below to show the useful energy transfer carried out by a train which runs on diesel.

.... energy	train →	.... energy

[2]

b Olumide has a toy wind-up train. He winds it up with the key in its side to make it move. When he winds it up he makes the spring inside into a source of stored energy. What is this stored energy called?

....

.... [1]

c The diagram represents the useful energy transfers in a car which make it move. Fill in the gaps using the correct words from the list.

thermal **chemical** **kinetic** **elastic** **sound**

.... energy	Car engine →	.... energy

[2]

d From the list above name one type of wasted energy to which energy from petrol is transferred in a car engine.

.... [1]

e The movement of a car drives an alternator in the engine which generates electric current for use in the car. The diagram below represents this.
Fill in the gaps using the correct words from the list.

thermal **chemical** **kinetic** **elastic** **sound**

.... energy	alternator →	.... energy

[2]

5 Some pupils did an experiment and plotted a graph of how a spring stretches as the load increases. The diagram below shows the experiment.

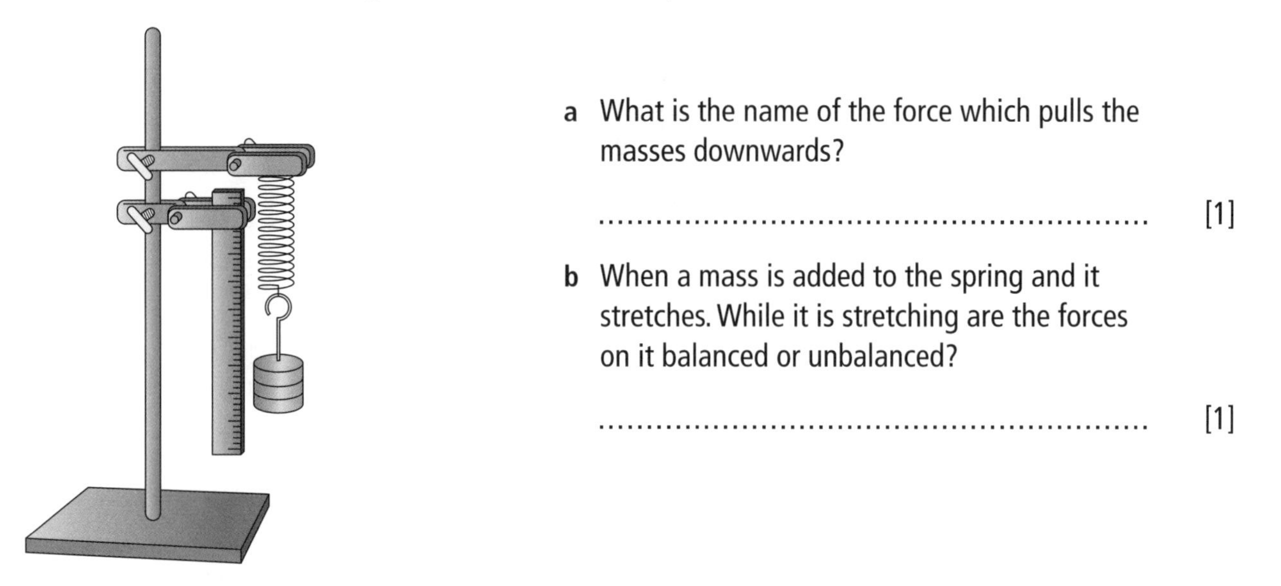

a What is the name of the force which pulls the masses downwards?

.... [1]

b When a mass is added to the spring and it stretches. While it is stretching are the forces on it balanced or unbalanced?

.... [1]

c The results of the experiment are recorded in the table below.

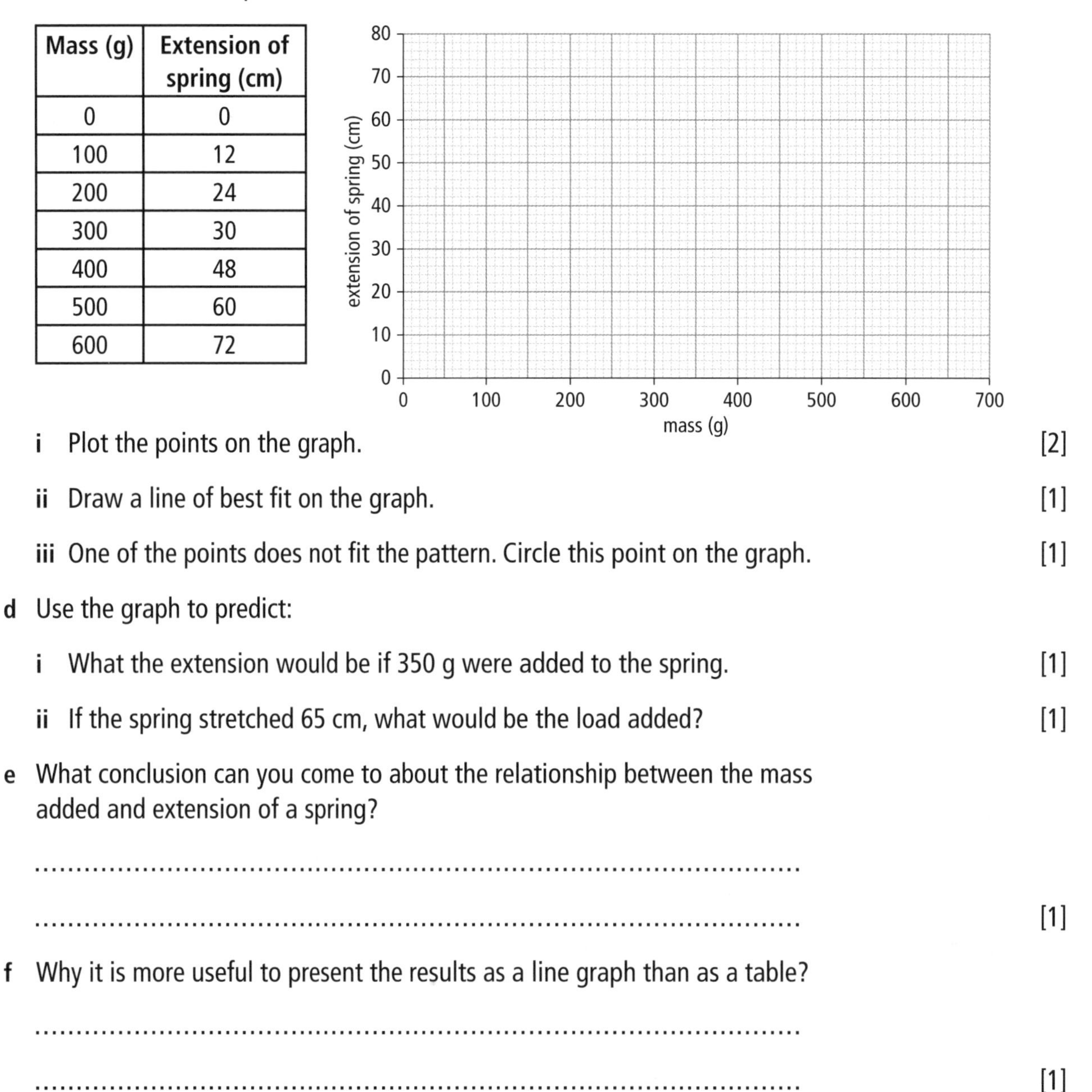

Mass (g)	Extension of spring (cm)
0	0
100	12
200	24
300	30
400	48
500	60
600	72

i Plot the points on the graph. [2]

ii Draw a line of best fit on the graph. [1]

iii One of the points does not fit the pattern. Circle this point on the graph. [1]

d Use the graph to predict:

i What the extension would be if 350 g were added to the spring. [1]

ii If the spring stretched 65 cm, what would be the load added? [1]

e What conclusion can you come to about the relationship between the mass added and extension of a spring?

....

.... [1]

f Why it is more useful to present the results as a line graph than as a table?

....

.... [1]

6 Kistna is experimenting with glass prisms. When he shines a light through one prism a spectrum with all the colours of the rainbow come out at the other side.

His theory is that the white light going into the prism is made up of these colours. He tries to find more evidence for his theory. He puts a second prism beside the first as shown in the diagram.

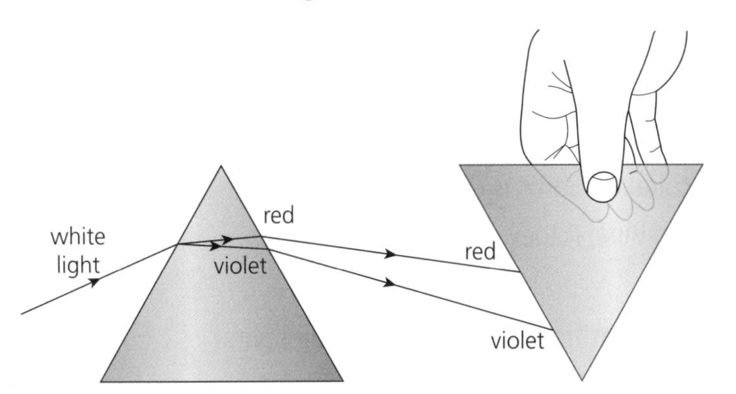

a i What happens to the spectrum of light as it passes through the second prism?

....

.... [2]

ii How does this provide evidence in support of Kistna's theory?

....

.... [1]

b Rainbows are spectrums formed by the sun shining through rain. What is the 'prism' which makes a rainbow?

....

.... [1]

The diagram shows how a prism breaks white light up into its colours.

c Explain what the lines in the diagram represent.

....

.... [1]

d Why do the lines marked blue bend more than the lines marked red?

....

.... [2]

e A theatre is putting on a play and need coloured lights on the stage. There are three spotlights – red, green, and blue. If they combine light from the red and green ones they get yellow.

i What colour do they get if they combine light from the blue and green lights?

.... [1]

ii How could they get white light from the three spotlights?

.... [1]

7 The bar chart shows the time taken for six planets to rotate once around their own axes. The time taken for one rotation of Uranus is 17 hours.

a On the chart, use a ruler to draw a bar for the planet Uranus.

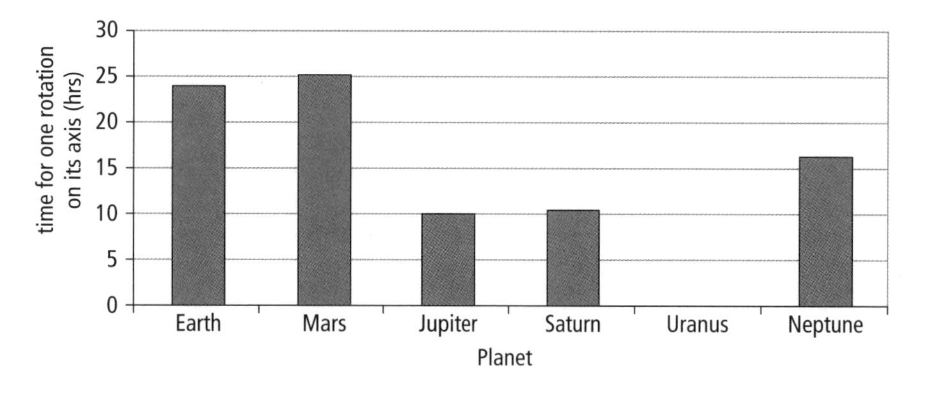

[1]

b The table below shows the temperatures on planets in the solar system.

Planet	Temperature on planet surface (°C)
Earth	22
Jupiter	–150
Mars	–23
Mercury	350
Neptune	–220
Saturn	–180
Uranus	–210
Venus	480

i Which planet's temperature is most similar to Earth's?

.... [1]

ii According to the information in the table which planet is the hottest?

.... [1]

iii According to the information in the table which planet is the coldest?

.... [1]

c In his science lesson Neela says that it is strange that Venus is hotter than Mercury. Explain Neela's comment.

....

.... [1]

d The universe contains planets, moons, stars, galaxies, and solar systems. Which is the largest?

....

.... [1]

e Fill in the correct units in this sentence:

Joel's mass is 65 .... His weight is 650 .... [2]

f The table shows what Joel's weight would be on some other planets.

Planet	Joel's weight
Mercury	234
Jupiter	1625
Saturn	728
Neptune	865

i On which planet in the list would he be heaviest?

.... [1]

ii Which planet in the list has the strongest force of gravity?

.... [1]

iii What would Joel's mass be on Mercury? Include the units.

.... [1]

8 Abasi is cycling home from school.

a The first part of his journey is along a clear flat cycle lane and he does 150 metres in 60 seconds. What is his speed?

.... [1]

b Abasi finds that if he lowers his head and body down as close to the handlebars as possible he can increase his speed on this stretch. Explain why.

.... [1]

c The graph below shows his speed along different stretches of his journey.

i Use the graph to calculate his average speed for the whole journey.

.... [1]

ii Along the way Abasi sees a friend and stops for a quick chat. Write down the letter of the part of the journey where he does this.

.... [1]

iii One part of the journey home has a steep hill which always slows him up. Write down the letter of the part of the journey where this happens.

.... [1]

9 Anyam completes an investigation into electromagnets. She wraps a piece of wire around a nail and picks up paperclips. She predicts that if there are more turns of wire then the electromagnet will be stronger.

These are the results of her investigation recorded in a table

Number of turns of wire	Number of paperclips picked up
0	0
20	1
40	2
60	6
80	10
100	12
120	15

a Plot the graph on the graph paper below.

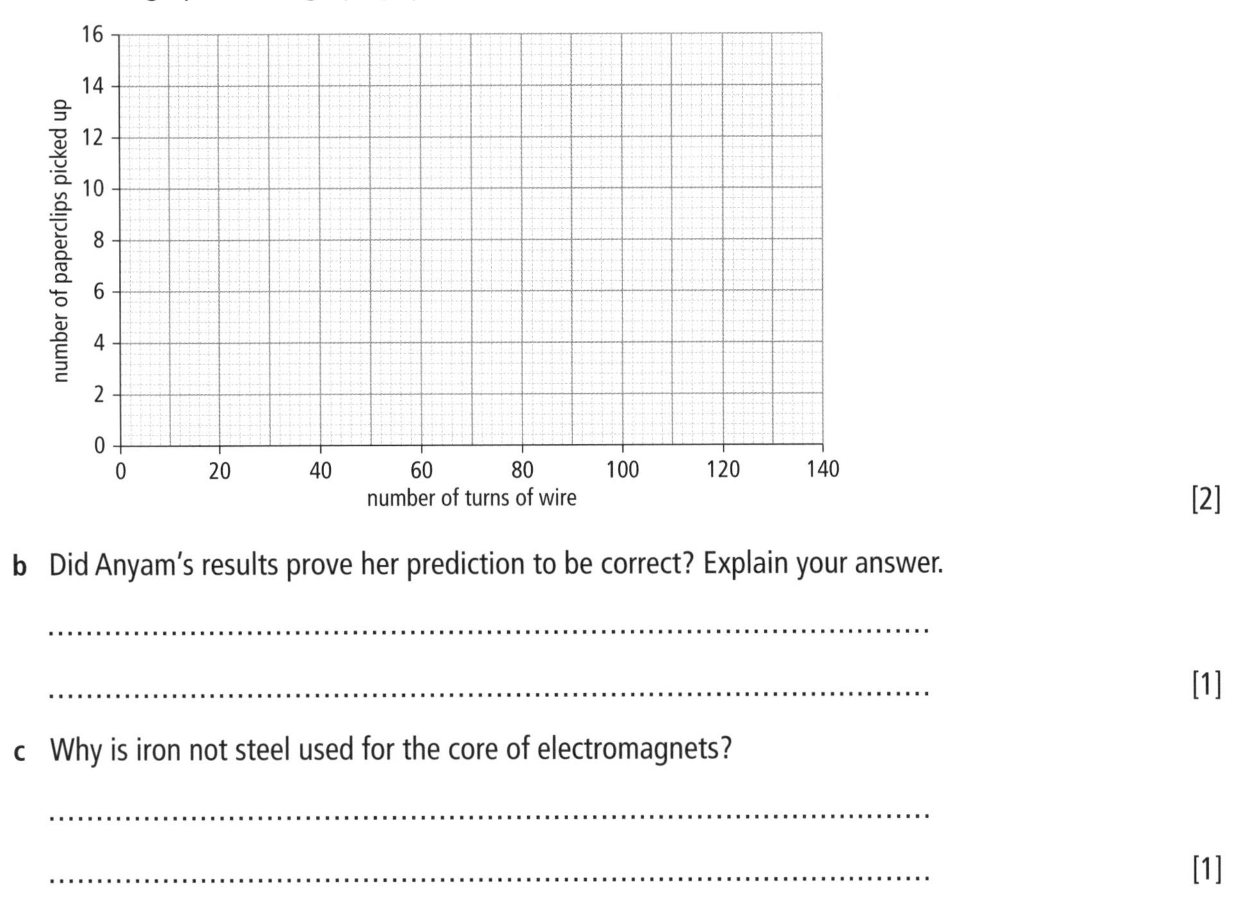

[2]

b Did Anyam's results prove her prediction to be correct? Explain your answer.

....

.... [1]

c Why is iron not steel used for the core of electromagnets?

....

.... [1]

10 Abi does an investigation to find the effect of the length of a pendulum on its period (time taken for one swing). She times one swing for several different lengths of string. She plots her results on the graph below.

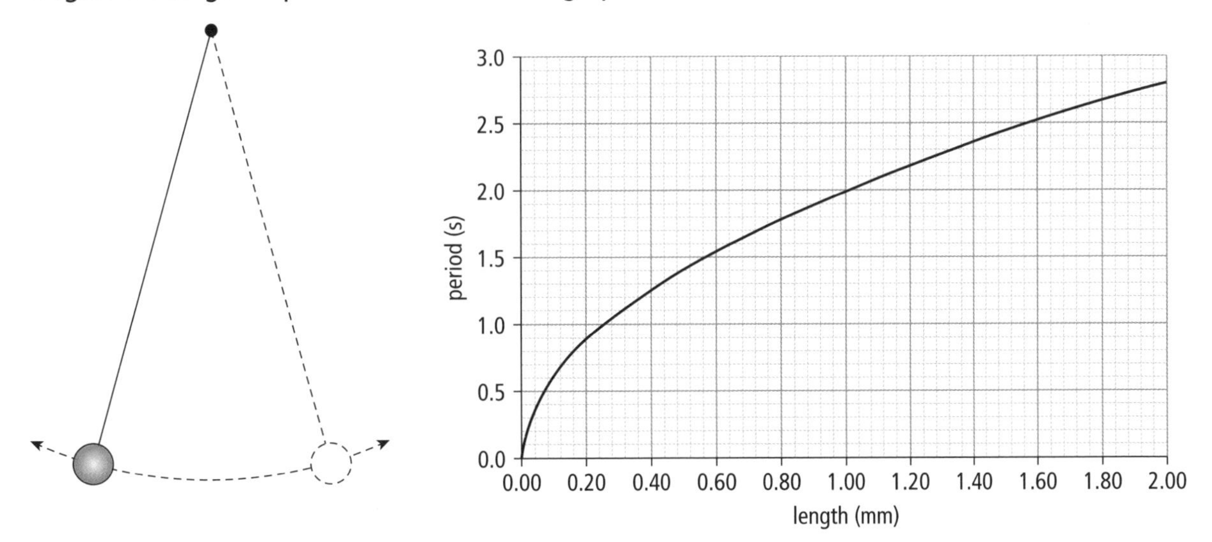

a Suggest two things which Abi should keep the same in her investigation.

....

.... [2]

b Suggest one way in which Abi could improve her investigation.

....

.... [1]

c Use the graph to predict how long the period would be at 1.1 metres.

....

.... [1]

d What conclusion would you draw from her graph show about the effect of length on the period of the pendulum?

....

.... [1]

11 Emeka is learning to play the guitar. To get the notes he presses down on the string of the guitar at its neck and plucks the string further down.

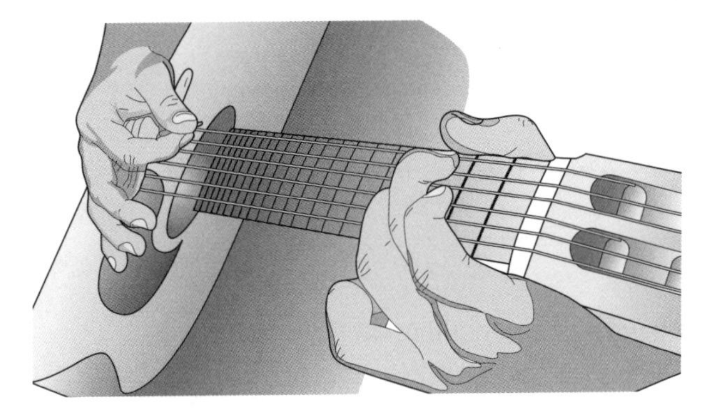

a How do the strings make a sound?

....

.... [1]

b What does the sound from the string travel through to reach Emeka's ear?

....

.... [1]

c Emeka moves his finger on the neck of the guitar slightly so that the string is shorter. What effect does this have on the sound? Tick the correct answer below.

The note is louder. ☐

The note is softer. ☐

The note is higher. ☐

The note is lower. ☐ [1]

d What happens to the note if Emeka plucks the string with more force?

....

.... [1]

Emeka decides to test the frequencies produced by a string of different lengths. His results were:

Length (mm)	Frequency (Hz)
100	3200
200	1600
400	800
800	400

e Predict what the frequency might be if the string was 1600 mm long.

....

.... [1]

f What is the relationship between string length and frequency?

....

.... [1]

g Describe two things they should keep constant in this experiment.

....

.... [2]

1 Use the words in the box below to complete the sentences which follow.

light	sound	fuel	thermal	kinetic	warm	electrical

a Light bulbs are useful for us because they change .... energy into .... energy.

b Radios which work from batteries transfer .... energy in the batteries into .... energy.

c Traditional light bulbs also give out a lot of .... energy which is less useful for us. [5]

2 A magnet is surrounded by a magnetic field.

a Which of the following diagrams shows the correct shape of a magnetic field around a bar magnet? Tick (✓) the correct box.

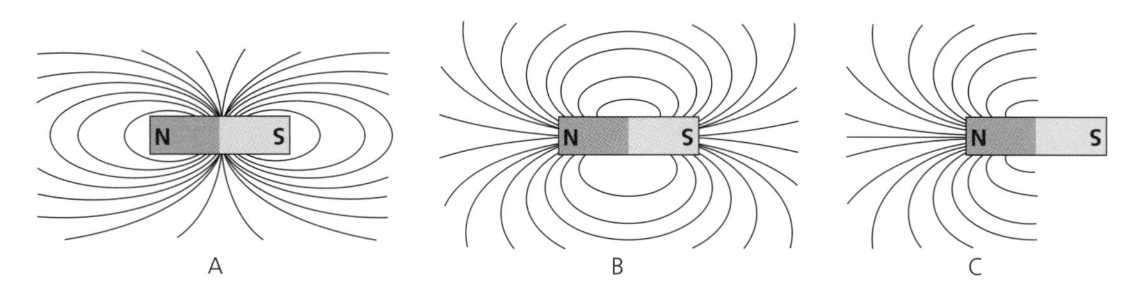

b The magnetic field around a magnet is invisible. Describe an experiment you could do to find its shape using iron filings.

....

.... [3]

3 The diagram shows four forces acting on a ship travelling across the Indian Ocean.

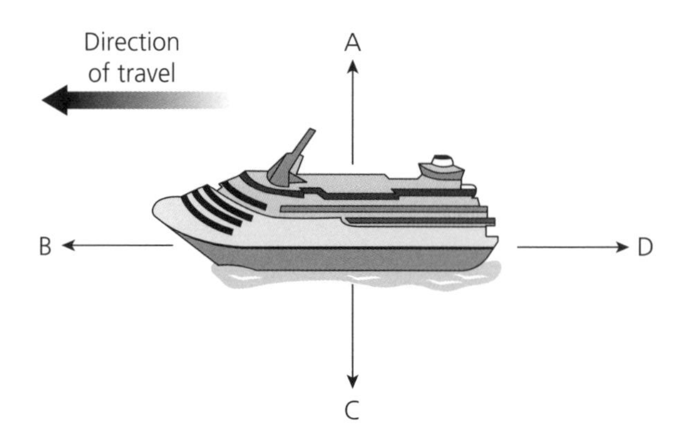

a Which arrow represents upthrust from the water? Write down the letter.

....

b Which **two** forces must be balanced, however fast the boat is travelling? Write down the letters.

.... and ....

c When the ship is travelling at a steady speed in the direction shown, which two forces are balanced?

....

d The ship speeds up. Which of the following statements is true? Tick (✓) the box next to the correct statement.

Force B is zero. ☐

Force B is greater than force D. ☐

Force D is equal to force B. ☐

Force D is greater than force B. ☐ [4]

4 The Earth's tilt on its axis as it orbits the Sun gives us the four seasons, summer, autumn, winter, and spring.

a Look at the tables below. Draw lines to match each season with the correct description of its day length and temperature. One has been done for you.

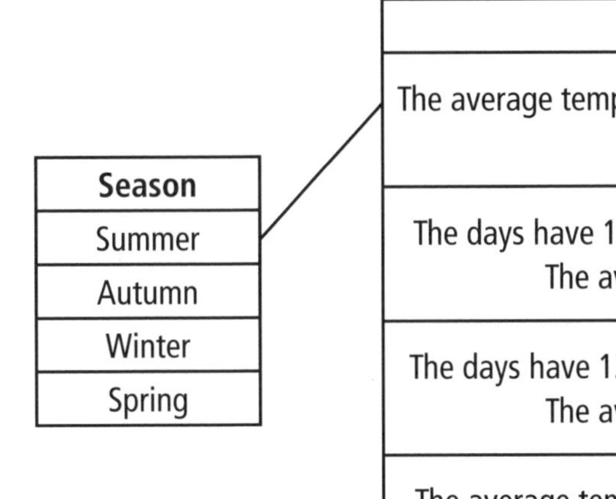

Season
Summer
Autumn
Winter
Spring

Description
The average temperature is 20 °C and there are 16 hours of sunlight in the day.
The days have 12 hours of light and are getting longer. The average temperature is 13 °C.
The days have 12 hours of light and are getting shorter. The average temperature is 13 °C.
The average temperature is 6 °C and there are 8 hours of light in the day.

b The diagrams below show how the Earth is tilted towards the Sun in summer and winter. Explain why this causes different lengths of day in the northern hemisphere.

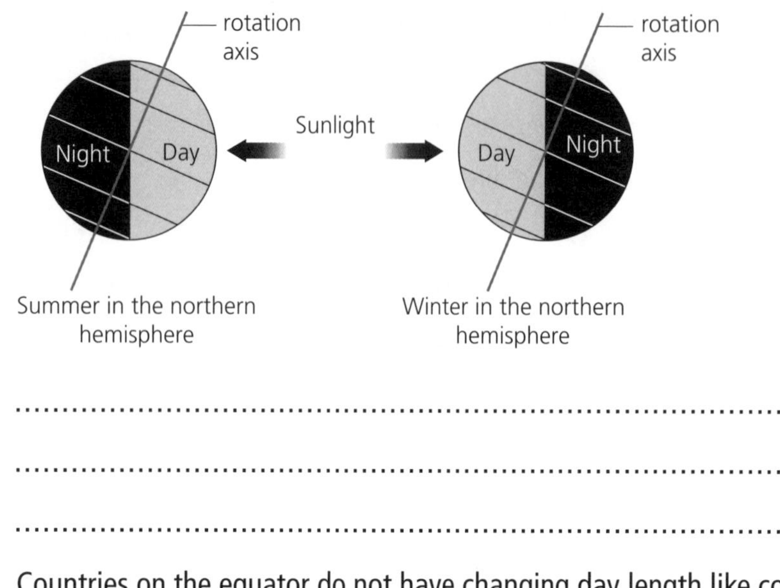

....

....

....

c Countries on the equator do not have changing day length like countries nearer the poles. Why?

....

....

.... [6]

5 Kiran and Laxmi were measuring the voltage at different points in the circuit they had set up.

a What is voltage? Tick the box (✓) next to the **two** correct statements below:

Voltage measures how much electrical energy is used by a component. ☐

Voltage measures the number of electrons flowing through a component. ☐

Voltage measures how much electrical energy a battery provides in a circuit. ☐

Voltage measures the number of electrons flowing from a battery into a circuit. ☐ [2]

b The diagram below shows the voltmeters they set up in their first investigation. They used two identical bulbs in their circuit.

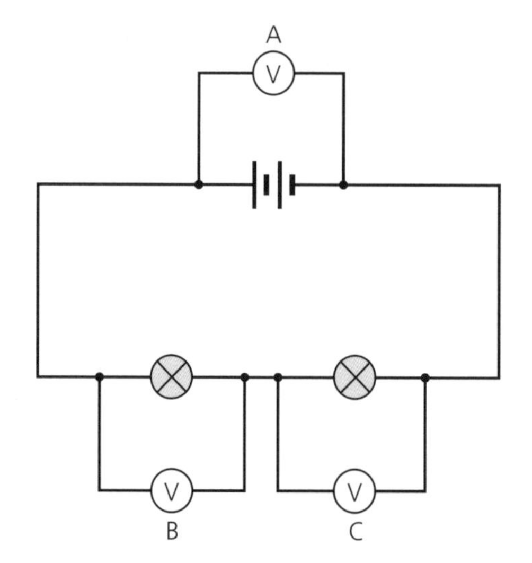

If the reading from the meter at B is 6 volts, what is the reading:

i at A? ....

ii at C? ....

c They then set up a different circuit as shown. If the voltage at A in this circuit is 6 volts, what will it be at B and C when the close the switch?

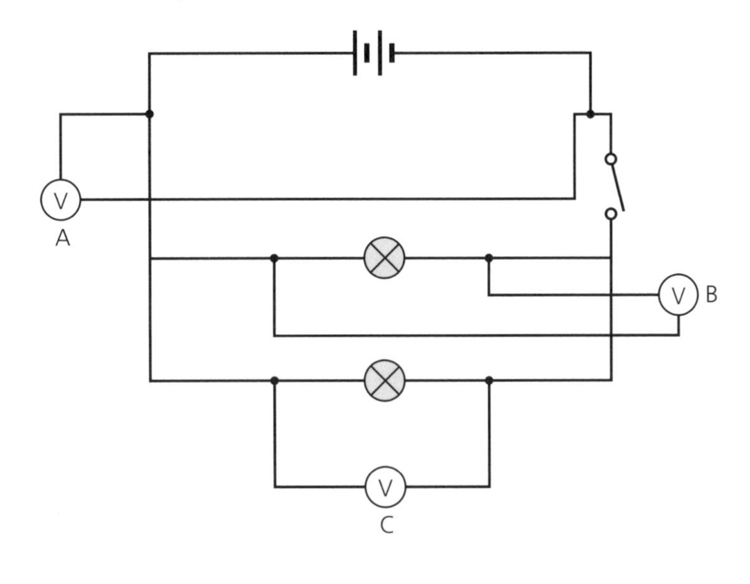

.... [2]

6 The graph below shows Anya's journey home from school.

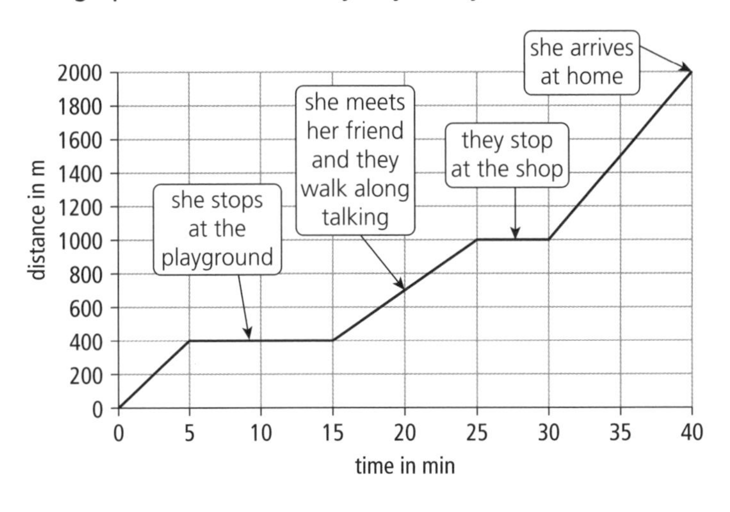

a How long does Anya stop at the playground?

....

b After they leave the shop Anya and her friend go in different directions. How far have they walked together?

....

c What is Anya's speed when she was walking with her friend?

....

d When they leave the shop Anya realises she is late so she walks faster. What is her speed for the last part of the journey from the shop to home?

....

e What is Anya's **average** speed for the whole walk home?

.... [5]

7 Lila and Navin set up the equipment shown below. They use blobs of wax which are each roughly the size of a pea.

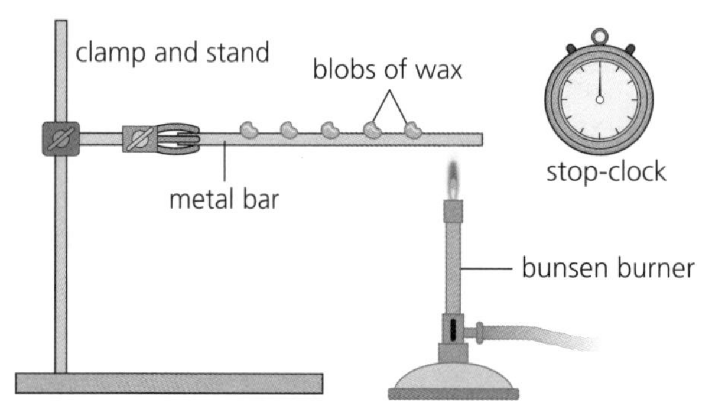

They time how long it takes for each blob of wax to melt.

a Describe one safety precaution that Emily and Cameron should take when carrying out their experiment.

....

b Lila and Navin record their results in the table shown below.

	Time for blob to melt (s)			
Distance to blob (cm)	**1**	**2**	**3**	**Average**
2	14	13	12	13
3	27	26	25	26
4	41	41	(52)	41
5	53	52	54	53

i Lila and Navin have circled one of their results. They did not include this result when calculating their averages. Explain why.

....

ii Suggest one thing in their procedure which might explain this result.

....

iii Give one way in which they could improve their experiment.

....

c i What type of heat transfer are Emily and Cameron investigating?

....

ii Using a particle model explain how heat is transferred from the end of the metal bar above the Bunsen burner to the other end.

....

.... [10]

8 The diagrams below show the traces made on an oscilloscope screen by sound waves from four different sounds.

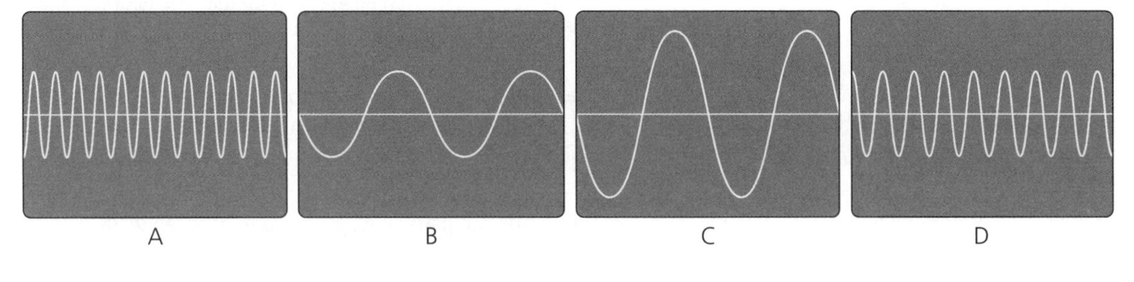

a i Which shows the highest frequency sound?

....

ii Explain what this means about this sound would sound like compared to the others.

....

....

b i Which shows the highest amplitude?

....

ii Explain what this means about this sound would sound like compared to the others.

....

....

c Which of these four sounds would transfer the most energy and why?

....

.... [6]

Absorbed energy of electromagnetic radiation (e.g. light) or sound is transferred to thermal energy on passing through or into a medium
Accelerate speed up, get faster
Acceleration the rate of change of increasing speed (the amount by which speed increases in one second)
Accuracy (of a measurement) how correct a measurement is – how close to its true value
Air resistance the force on an object that is moving through the air, causing it to slow down (also known as drag)
Alloy a material made of a mixture of metals, or of carbon with a metal
Ammeter a device for measuring electric current in a circuit
Ampere (amp) the unit of measurement of electric current, symbol A
Amplifier a device for making a sound louder
Amplitude the distance from the middle to the top or bottom of a wave
Analogy a way of explaining something by saying that it is like something else
Andromeda the nearest galaxy to the Milky Way
Angle of incidence the angle between the incident ray and the normal line
Angle of reflection the angle between the reflected ray and the normal line
Angle of refraction the angle between the refracted ray and the normal line
Anomalous point a point on a graph that does not fit the general pattern (also called an outlier)
Anticlockwise the direction of rotation that is opposite to the movement of the hands of a clock
Apparent depth how deep something underwater appears to be when viewed from above
Archimedes' principle a law that says the upthrust on an object is equal to the weight of water displaced
Armature the coil of wire in an electromagnetic device such as a generator
Artificial satellites spacecraft made by people to orbit the Earth for various purposes
Asteroid a lump of rock in orbit around the Sun
Asteroid belt a large number of asteroids between Mars and Jupiter
Astronomer a scientist who studies space
Atmosphere the layer of air above Earth's surface
Atmospheric pressure the force on an area of the Earth's surface due to the weight of air above it, or the pressure in the atmosphere
Atom the smallest particle of an element that can exist
Attract pull together, for example opposite poles of a magnet or positive and negative charges attract each other
Audible can be heard
Auditory canal the passage in the ear from the outer ear to the eardrum
Average found by adding a set of values together and dividing by the number of values (also called the mean)
Average speed the total distance travelled divided by the total time taken for a complete journey
Average speed (molecules) the typical value of the speed of molecules in a gas or liquid
Axis, of Earth the imaginary line through the Earth around which it spins
Balanced forces describes forces that are the same size but act in opposite directions on an object
Bar chart a way of presenting data in which only one variable is a number
Battery two or more electrical cells joined together
Big Bang the expansion of space which we believe started the Universe
Biodiesel a biofuel made from plant oils
Bioethanol a biofuel made from carbohydrates such as sugar
Biofuel a fuel produced from renewable resources
Biogas gas produced from waste products, usually methane, used to generate electricity
Biomass material from plants used for fuel, e.g. wood
Black dwarf a remnant of a star like our Sun that no longer gives out light
Black hole a remnant of a star much bigger than our Sun from which nothing can escape, not even light
Boil to change from a liquid into a gas at the boiling point

Capacitor two plates with an insulator between them, used to store charge
Carbon dioxide a gas found in small amounts in the atmosphere which plants use to make food and which is a greenhouse gas
Categoric describes a variable whose values are words not numbers
Cell (electrical) a device that is a store of chemical energy which is transferred to electrical energy in a circuit
Centre of gravity the point in an object where the force of gravity acts
Centre of mass the point in an object where the mass appears to be concentrated
Centripetal force the force directed towards the centre that causes a body to move in a uniform circular path
Charge positive or negative, a property of protons and electrons
Charge-coupled device (CCD) a grid of components that work like capacitors at the back of a digital camera, and produce a digital signal
Chemical potential energy (chemical energy) energy stored in fuels, food, and electrical batteries
Chemical reaction an event that creates new substances
Circuit (electric) a complete pathway for an electric current to flow
Circuit diagram a way of showing a circuit clearly, using symbols
Circuit symbol a drawing that represents a component in a circuit
Climate change changes to long-term weather patterns as a result of global warming
Clockwise the direction of rotation that is the same as the movement of the hands of a clock
Coal a fossil fuel formed from dead plants that have been buried underground over millions of years
Coal-fired power station a place where the fossil fuel coal is burned to generate electricity
Cobalt a magnetic material
Cochlea a snail-shaped tube in the inner ear where the sensory cells that detect sound are
Colour blindness someone with colour blindness cannot tell certain colours apart, because some cone cells in the retina of the eye do not work properly
Comets bodies in space made of dust particles frozen in ice, which orbit the Sun

Communicate to share and exchange information
Compact disc a metal disc that can store high-quality digital recordings
Compass a device containing a small magnet that is used for finding directions
Component an item used in an electric circuit, such as a lamp
Compressed squashed into a smaller space
Compression the part of a sound wave where the air particles are close together
Conclude to make a statement about what the results of an investigation tell you
Conduction (of charge) the movement of charge through a material such as metal or graphite, forming an electric current
Conduction (of thermal energy) the way in which thermal energy is transferred through solids (and to a much lesser extent in liquids and gases)
Conductor a material such as a metal or graphite that conducts charge or thermal energy well
Cone a specialised cell in the retina that is sensitive to bright light and colour
Consequence (risk) what can happen as a result of something you do
Conservation (of energy) a law that says that energy is never made or lost but is always transferred from one form to another
Constant not changing
Constellation a collection of stars that make a pattern in the sky
Contact force a force that acts when an object is in contact with a surface, air, or water
Continuous variable a variable that can have any value across a range, such as time, temperature, length
Convection the transfer of thermal energy by the movement of a gas or liquid
Convection current the way in which thermal energy is transferred through liquids and gases by the movement of their particles
Conventional current the current that flows from the positive to the negative terminal of a battery
Core a rod of a magnetic material placed inside a solenoid to make the magnetic field of an electromagnet stronger
Cornea the transparent layer at the front of the eye
Correlation a link between two things; it does not necessarily mean that one thing causes the other
Creative thinking thinking in a new way
Critical angle the smallest angle of incidence at which total internal reflection occurs
Crude oil a thick black liquid formed underground from the remains of prehistoric plants and animals that died millions of years ago. It is used to make fuels such as petrol/gasoline and diesel, and many plastics.
Current the flow of electric charge (electrons) around a complete circuit

Data measurements taken from an investigation or experiment
Day the period of time when one section of the Earth (or other planet) is facing the Sun
Decelerate slow down, get slower
Deceleration the amount by which speed decreases in one second
Decibel (dB) a commonly used unit of sound intensity or loudness
Deform change shape
Degrees Celsius (°C) a temperature scale with 0 °C fixed at the melting point of ice and 100 °C fixed at the boiling point of water
Demagnetise to destroy a magnet by heating it up, hitting it, or putting it in an alternating current
Density the mass of a substance in a certain volume
Dependent variable the variable that changes when you change the independent variable
Detector something that absorbs electromagnetic radiation or sound to produce a signal
Dielectric the insulating material between the plates of a capacitor
Diffuse particles in gases and liquids move from where there are a lot of the particles to where there are fewer
Diffuse describes reflection from a rough surface
Diffusion the way particles in liquids and gases mix or spread out by themselves
Digital signal a signal that is used to transfer information between sensors and computers
Directly proportional a relationship in which one quantity increases in the same way as another
Discrete describes a variable that can only have whole-number values
Dispersion the splitting up of a ray of light of mixed wavelengths by refraction into its components
Distance multiplier a type of lever that uses a larger force to produce a smaller force at a larger distance
Distance–time graph a graph showing how the distance travelled varies with time
Domain a small region inside a magnetic material that behaves like a tiny magnet
Drag a force on an object moving through air or water, causing it to slow down
Dwarf planet a lump of rock in orbit around the Sun that is nearly spherical but has other objects around it
Dynamo a device that transforms kinetic energy into electrical energy (a small generator)

Ear defenders a device used to protect the ears from noise
Eardrum a membrane that transmits sound vibrations from the outer ear to the middle ear
Earth a rocky inner planet, the third planet from the Sun
Earth (charge) to connect a metal wire from an object to the ground to take any charge away
Earthing the process of connecting objects to the ground
Echo a reflection of a sound wave by an object
Echolocation the process of finding out where something is using echoes
Eclipse the Sun or Moon is blocked from view on Earth (see also lunar eclipse or solar eclipse)
Efficient describes something that does not waste much energy
Effort the amount of force that you use to push down when using a lever
Elastic describes a type of material that can be stretched and will return to its original length when the pulling force is removed
Elastic limit the point beyond which a spring will never return to its original length when the pulling force is removed
Elastic potential energy (EPE) energy stored in an elastic object that is stretched or squashed
Electric car car powered by electric batteries
Electric circuit a complete pathway for an electric current to flow
Electric current a flow of electric charge (electrons) around a complete circuit
Electrical energy energy that is transferred from a cell or battery to the components of a circuit. A light bulb changes electrical energy to light and thermal energy.

Electrical signal (ear) information is transferred from the ear to the brain as an electrical signal (nerve signal)
Electromagnet a temporary magnet produced using an electric current
Electromagnetic radiation radiation with electric and magnetic properties that can travel through a vacuum (such as the Sun's radiation)
Electromagnetic spectrum the range of wavelengths of electromagnetic radiation produced by the Sun and other sources
Electron tiny charged particle in an atom, that flows through a wire to create an electric current
Electrostatic force the force between two charged objects
Electrostatic phenomena things that happen because objects have become charged
Element a substance consisting of atoms of only one type
Emit to send something out (such as heat, light, vapour)
Endoscope a medical instrument for seeing inside the human body
Energy this is needed to make things happen
Energy conservation energy is never made or lost but may be transferred from one form to another, although they are not always forms we can use
Energy converter something that produces a secondary source of energy, such as a power station
Energy transfer energy changing from one form to another, such as from chemical to thermal energy
Equator an imaginary line round the middle of the Earth at an equal distance from both the North and South Poles
Equilibrium balanced (as in a lever or see-saw)
Evaporate to turn from a liquid to a vapour (gas)
Evaporation the process of turning from a liquid to a gas without boiling
Evidence observations and measurements that support or disprove a scientific theory
Exojoule a million, million, million joules
Exoplanet a planet in orbit around a star other than our Sun
Expand to increase in size, get bigger
Explanation a statement that gives a reason for something using scientific knowledge
Extension the amount an object gets longer when you stretch it
Eye the organ of sight, which focuses and detects light

Filament the very thin, coiled piece of wire inside a light bulb that glows
Filter a piece of material that allows some radiation (colours) through but absorbs the rest
Floating an object floats when the upthrust from water is equal to the downwards force of the object's weight
Force a push or a pull that acts on an object to affect its movement or shape
Force multiplier a lever or hydraulic machine that can lift or move heavy weights using a force smaller than the weight
Forcemeter a device used to measure forces
Fossil fuel fuels made from the decayed remains of animals and plants that died millions of years ago. Fossil fuels include coal, oil, and natural gas.
Frequency the number of complete waves or vibrations produced in one second (measured in hertz)
Friction a force that resists movement because of contact between surfaces
Fuel a material that contains a store of energy and can be burned, e.g. gas, oil, coal, petrol/gasoline
Fuel cell a device that uses chemical reactions to generate electricity
Fulcrum the point about which a lever or see-saw turns
Fundamental (sound) the lowest frequency of sound

Galaxy a number of stars and the solar systems around them grouped together
Gas (natural) a fossil fuel that collects above oil deposits underground
Gas pressure (air pressure) the force exerted by air particles when they collide with 1 square metre (1 m^2) of a surface
Generator a device that uses kinetic energy to induce a voltage
Geocentric model a model of the Universe with the Earth at the centre
Geothermal an energy source that uses the thermal energy underground to produce electricity
Global positioning system (GPS) a system that pinpoints the position of something using signals from a satellite
Gravitational field a region in which there is a force on a mass due to its attraction to another mass
Gravitational field strength the force on a mass of 1 kg, measured in N/kg
Gravitational force (gravity) the force of attraction between two objects because of their mass
Gravitational potential energy (GPE) energy stored in an object because of its height above the ground
Greenhouse gases gases that contribute to global warming, such as carbon dioxide, water vapour, and methane

Harmonics frequencies of a sound wave that are multiples of the fundamental frequency
Hazard symbols warning symbols used on chemicals which show what harm they might cause if not handled properly
Heat to change the temperature of something; the word 'heat' is sometimes used instead of thermal energy
Heat pump a device that transfers heat from the ground to a building on the surface
Heliocentric model a model of the Universe with the Sun at the centre
Hertz (Hz) the unit of frequency
Hubble space telescope a telescope orbiting the Earth and sending back clear pictures of space
Hybrid car a car that can run either on electricity from a battery or on petrol/gasoline in a petrol engine
Hydraulic brakes brakes that use a liquid in pipes to transfer forces and make them bigger
Hydraulic machine a machine that uses a liquid in pipes to transfer forces and make them bigger
Hydraulic press a press that uses a liquid in pipes to transfer forces and make them bigger
Hydroelectricity electricity generated using the energy of water falling downhill
Hydrogen a non-metal element that exists as a gas at everyday temperatures

Image the point from which rays of light entering the eye appear to have originated
Incident ray the ray coming from a source of light
Incompressible describes something that cannot be compressed (squashed)
Independent variable the variable that you change, that causes changes in the dependent variable
Induced (voltage) a voltage produced when a conductor is in a changing magnetic field

Inertia the tendency of an object to resist a change in speed caused by a force
Infinite without end
Infrared (radiation) a type of electromagnetic radiation that transfers thermal energy from a hotter to a colder place, also known as heat
Inner ear the part of the ear made up of the cochlea and semi-circular canals
Inner planets Mercury, Venus, Earth, and Mars
Insulator a material that does not conduct thermal energy or electricity very well
Intensity (sound) how loud a sound is, measured in decibels
International Space Station (ISS) a research station in orbit around the Earth
Interstellar space the space between stars or solar systems
Inversely proportional a relationship in which one quantity decreases as the other increases
Inverted upside down
Investigation an activity such as an experiment or set of experiments designed to produce data to answer a scientific question or test a theory
Iron a metal element that is the main substance in steel. Iron is a magnetic material.
Iron core an iron rod placed in a coil to increase the magnetic field strength when a current flows in the coil

Joule the unit of energy, symbol J
Jupiter a large outer planet made of gas, fifth from the Sun

Kaleidoscope a toy containing mirrors and coloured glass or paper
Kilogram the unit of mass, symbol kg
Kilojoule (kJ) 1000 joules
Kilometre per hour the unit of speed, km/h
Kilowatt 1000 watts
Kinetic energy movement energy
Kuiper belt the region outside the Solar System where astronomers think that some comets come from

Laser a device that produces an intense beam of light that does not spread out
Laterally inverted the type of reversal that occurs with an image formed by a plane mirror
Law of conservation of energy the law that says that energy cannot be created or destroyed
Law of reflection the law that says that the angle of incidence is equal to the angle of reflection
Lens a device made of shaped glass which focuses light rays from objects to form an image
Lever a simple machine consisting of a rigid bar supported at a point along its length
Life cycle (of a star) the process that describes how a star is formed and what will happen to it
Light a form of electromagnetic radiation that comes from sources like the Sun
Light-emitting diode (LED) a low-energy lamp
Light energy energy transferred by sources such as the Sun and light bulbs
Light intensity the energy per square metre, measured in lux
Light sources objects that emit visible light, also called luminous objects
Light year the distance light travels in one year
Lightning conductor a piece of metal connected to tall buildings to conduct lightning to the ground
Line graph a way of presenting results when there are two numerical variables
Line of best fit a smooth line on a graph that travels through or very close to as many of the points plotted as possible
Liquid pressure the pressure produced by collisions of particles in a liquid
Load an external force that acts over a region of length, surface, or area
Lodestone a naturally occurring magnetic rock
Longitudinal describes a wave in which the vibrations are in the same direction as the direction the wave moves
Loudspeaker a device that changes an electrical signal into a sound wave
Lubrication reducing friction between surfaces when they rub together
Luminous describes something that gives out light
Lux the unit of light intensity

Mach the ratio of the relative speed to the speed of sound
Magnet an object that that attracts magnetic materials and repels other magnets
Magnetic field an area around a magnet where there is a force on a magnetic material or another magnet
Magnetic field lines imaginary lines that show the direction of the force on a magnetic material in the magnetic field
Magnetic force the force between the poles of two magnets, or between a magnet and a magnetic material such as iron
Magnetic material a material that is attracted to a magnet, such as iron, steel, nickel, or cobalt
Magnetic resonance imaging (MRI) scanner a machine that uses strong magnetic fields to produce images of the inside of the human body
Magnetised made into a magnet
Magnetism the property of attracting or repelling magnets or magnetic materials
Main sequence the longest stage of a star's life cycle; the current stage of our Sun
Mains electricity (mains supply) electricity generated in power stations and available through power sockets in buildings
Mars a rocky planet, fourth from the Sun
Mass the amount of matter in an object. The mass affects the acceleration for a particular force.
Measuring cylinder a cylinder used to measure the volume of a liquid
Medium (sound/light) the material that affects light or sound by slowing it down or transferring the wave
Meniscus the curved upper surface of a liquid
Mercury the rocky inner planet nearest the Sun
Meteor a piece of rock or dust that makes a streak of light in the night sky
Meteorite a stony or metallic object that has fallen to Earth from outer space without burning up
Metre per second the unit of speed, m/s
Microphone a device for converting sound into an electrical signal
Middle ear the eardrum and ossicles (small bones) that transfer vibrations from the outer ear to the inner ear
Milky Way the galaxy containing our Sun and Solar System
Milliamp one thousandth of an amp
Minerals chemicals in rocks
Model a simplified description of a process. A model may be a physical model built on a different scale to the original system, or it may take the form of equations.

Moment a measure of the ability of a force to rotate an object about a pivot
Monochromatic describes light of a single colour or wavelength
Moon a rocky body orbiting Earth; it is Earth's only natural satellite
Moons the natural satellites of planets
Movement energy the energy of movement, also called kinetic energy

Natural satellite a moon in orbit around a planet
Nebula a region of dust and gas where stars are born
Negative describes the charge on an electron, or the charge on an object that has had electrons transferred to it
Negatively charged describes an object that has had electrons transferred to it
Neptune a large outer planet made of gas, eighth from the Sun
Neutral describes an object that has no charge; its positive and negative charges cancel out
Neutral point (magnetic field) a point where there is no force on a magnet or magnetic material because two or more magnetic fields cancel out
Neutralise to cancel out, when you add an equal amount of positive charge to negative charge
Newton the unit of force including weight, symbol N
Newtonmetre the unit of moment, symbol Nm
Nickel a magnetic material
Night the period on one section of the Earth or other planet when it is facing away from the Sun
Noise any undesired or unwanted sound
Non-contact force a magnetic, electrostatic, or gravitational force that acts without being in contact with something
Non-luminous describes objects that produce no light; objects that are seen by reflected light
Non-renewable describes an energy source that will run out eventually (such as fossil fuels)
Normal an imaginary line at right angles to a surface where a light ray strikes it
Normal brightness standard brightness of a single bulb lit by a single cell
North pole the pole of a magnet that points north. A north pole repels another north pole.
Northern hemisphere the half of the Earth between the equator and the North Pole
Nuclear energy the energy from nuclear fusion that powers the Sun and stars, or from uranium in nuclear power stations
Nuclear fusion the process of joining hydrogen together in the Sun and other stars that releases energy
Nucleus the centre of an atom that contains neutrons and protons

Object something that can be seen or touched
Observations the results of looking carefully at something and noticing properties or changes
Oil a fossil fuel formed from sea creatures over millions of years
Oort cloud a cloud of comets and dust outside the Solar System
Opaque describes objects that absorb, scatter, or reflect light and do not allow any light to pass through
Optic nerve a sensory nerve that runs from the eye to the brain
Optical fibres a very fine tube of plastic or glass that uses total internal reflection to transmit light
Orbit the path taken by one body in space around another (such as the Earth around the Sun)
Oscilloscope a device that enables you to see electrical signals that change, such as those made in a microphone
Ossicles the small bones of the middle ear (hammer, anvil, and stirrup) that transfer vibrations from the eardrum to the oval window
Outer ear the pinna and auditory canal
Outer planets Jupiter, Saturn, Uranus, and Neptune
Oval window the membrane that connects the ossicles to the cochlea in the ear
Oxide a compound made when an element combines with oxygen

Parallel circuit an electric circuit in which there are two or more paths for an electric current
Particles tiny pieces of matter
Pascal the unit of pressure, symbol Pa, equal to 1 N/m^2
Payback time the time taken to recoup the cost of something, such as home insulation or a wind turbine
Pendulum any rigid body that swings about a fixed point
Penumbra the area of blurred or fuzzy shadow around the edges of the umbra
Perforate to make a hole in something
Period the time taken to complete one cycle of motion
Periscope a tube with mirrors or prisms that enables you to see over objects
Permanent magnet a piece of metal that stays magnetic
Permanently extended the irreversible extension of a spring when loaded beyond its elastic limit
Petrol (gas or gasoline) a hydrocarbon fuel (containing hydrogen and carbon) that comes from crude oil
Phases of the Moon parts of the Moon we see as it orbits the Earth
Photosynthesis the process by which green plants make their own food from carbon dioxide and water using solar energy
Pie chart a way of presenting data in which only one variable is a number
Pinna the outside part of the ear that we can see
Pitch a property of sound determined by its frequency
Pivot a support on which a lever turns or oscillates
Pixel the small square that digital images are made of; a picture element
Plane mirror a mirror with a flat reflective surface
Planet any large body that orbits a star in a solar system
Plastic a type of material that can be stretched and does not return to its original length
Pluto used to be regarded as the ninth and last planet from the Sun; now called a dwarf planet together with others of the same size that are beyond Pluto's orbit
Poles, of Earth the north and south points of the Earth connected by its axis of tilt
Poles, of magnet the opposite and most strongly attractive parts of a magnet
Positive (charge) describes the charge on a proton, or the charge on an object that has had electrons transferred away from it
Positively charged describes an object that has had electrons transferred away from it
Potential energy stored energy
Power the rate of transfer of energy, measured in watts
Power station a place where fuel is burned to produce electricity
Precision the number of decimal places given for a measurement
Prediction statement saying to say what you think will happen
Preliminary work the work that you do before or during the planning stage of an investigation, to work out how to do it
Pressure the force applied by an object or fluid divided by the area of surface over which it acts
Pressure gauge an instrument for measuring pressure in a liquid or gas

Primary colours of light are red, blue, and green
Primary data data collected directly by scientists during a particular investigation
Primary sources (of energy) energy sources from the environment or underground, such as coal, uranium, or the wind
Principle of moments law that says that the clockwise moments equal the anticlockwise moments
Prism a triangular-shaped piece of glass used to produce a spectrum of light
Probability (risk) the chance that something will happen
Property a characteristic, for example wavelength and amplitude are properties of a wave
Proportional a relationship in which two variables increase at the same rate, for example when one is doubled the other doubles too
Proxima Centauri the nearest star to our Sun
Pupil the hole in the front of your eye where the light goes in

Rainbow an optical phenomenon that appears as the colours of the spectrum when falling water droplets are illuminated by sunlight
Rarefaction the part of a sound wave where the air particles are most spread out
Ray diagram a model of what happens to light shown by drawing selecting rays
Reaction time in humans, the time the brain takes to process information and act in response to it
Real describes an image that you can put on a screen, or the image formed in your eyes
Real depth the depth underwater that an object actually is
Receiver (sonar) a device that absorbs sound waves
Red giant part of the life cycle of a star like our Sun when it becomes much bigger and cooler
Reed switch a switch that uses a magnet to work
Refinery a place where crude oil is refined and separated into fuels
Reflected ray the ray that is reflected from a surface
Reflection the change in direction of a light ray or sound wave after it hits a surface and bounces off
Refract to change direction because of a change in speed
Refraction the change in direction of a light ray as a result of its change in speed
Refractive index a measure of how much light slows down in a medium
Refrigerant a liquid used in a refrigerator
Refrigerator a machine for keeping things cold using evaporation
Relay electrical device such that allows current flowing through it in one circuit to switch on and off a larger current in a second circuit
Reliable describes an investigation in which very similar data would be collected if it was repeated under the same conditions
Renewable describes energy resources that are constantly being replaced and are not used up, such as falling water or wind power
Repel to push away
Reservoir a large amount of water behind a dam used in hydroelectric power
Resistance how difficult it is for current to flow through a component in a circuit
Resultant force the single force equivalent to two or more forces acting on an object
Retina the layer of light-sensitive cells at the back of the eye
Reverberation the persistence of a sound for a longer period than normal
Risk a combination of the probability that something will happen and the consequence if it did
Rod a specialised cell in the retina that is sensitive to dim light

Sankey diagram a diagram that shows all the energy transfers taking place in a process, and the amount of energy in each transfer
Satellite any body that orbits another (such as the Moon or a weather satellite around Earth)
Saturn a large outer planet made of gas, sixth from the Sun
Scatter plot a graph that shows all the values in a set of measurements
Seasons changes in the climate during the year as the Earth moves around its orbit
Secondary colours colours that can be obtained by mixing two primary colours
Secondary data data collected by other scientists and published
Secondary sources (of energy) sources of energy that are produced from primary sources, such as electricity produced from coal, or petrol/gasoline produced from crude oil
Semicircular canals the part of the ear that helps you to balance
Series circuit an electrical circuit in which the components are joined in a single loop
Shadow an area of darkness on a surface produced when an opaque object blocks out light
Shield to put something in between a source and a receiver, for example sound is shielded by ear defenders
Signal (electrical) a voltage that changes over time
Solar cells devices that change light energy into electrical energy
Solar energy energy from the Sun which can be used directly to heat water, or to make electricity
Solar panels devices that transform light energy from the Sun into thermal energy
Solar System the Sun (our star) and the planets and other bodies in orbit around it. There are other solar systems in the Universe as well as our own
Sonar a system that uses ultrasound to detect underwater objects or to determine the depth of the water
Sound energy energy produced by vibrating objects
Sound level meter a device for measuring the intensity (loudness) of a sound
Sound wave a series of compressions and rarefactions that moves through a medium
Source (of light/sound) something that emits (gives out) light or sound
South pole the pole of a magnet that points south. A south pole attracts a north pole.
Southern hemisphere the half of the Earth between the equator and the South Pole
Spark a flash of light that you see when the air conducts electricity
Spectrum a band of colours produced when light is spread out by a prism
Speed the distance travelled in a given time, usually measured in metres per second, m/s
Speed of light the distance light travels in one second (300 million m/s)
Speed–time graph a graph that shows how the speed of an object varies with time

Spring balance a device for measuring forces, sometimes called a forcemeter or a newton meter
Spring a metal wire wound into spirals that can store elastic potential energy
Stable describes an object in equilibrium that cannot be easily toppled
Star a body in space that gives out its own light. The Sun is a star.
Steady speed a speed that doesn't change
Steel an alloy of iron with carbon and other elements. Steel is a magnetic material.
Streamlining describing a shape designed to reduce resistance to motion from air or liquid
Stretch the extension when an elastic material such as a spring is pulled outwards or downwards
Sun star at the centre of our Solar System
Sunlight light from the Sun
Sunspots dark spots on the surface of the Sun
Supersonic describes a speed that is faster than the speed of sound
Symbol a sign that represents something (see also circuit symbols, and hazard symbols)

Tangent a straight line that touches a curve or circle
Telescope a device made with lenses that allows distant objects to be seen clearly
Temperature a measure of how hot something is
Tension a stretching force
Terminal (of a cell) the positive or negative end of a cell or battery
Terminal velocity the highest velocity an object reaches when moving through a gas or a liquid; it happens when the drag force equals the forward or gravitational force
The bends the sickness that divers can suffer due to dissolved gases in their blood
Thermal energy the energy due to the motion of particles in a solid, liquid, or gas
Thermal image an image made using thermal or infrared radiation
thermal imaging camera a device that forms an image using thermal or infrared radiation so that different temperatures appear as different colours
Thermal a rising current of heated air
Thermometer a device used to measure temperature
Thrust force from an engine or rocket
Tidal energy/power energy from the movement of water in tides which can be used to generate electricity
Timbre the quality of a sound resulting from the harmonics present in the sound
Timing gates two sensors connected together used to measure speed or acceleration precisely and accurately
Total internal reflection the complete reflection of light at a boundary between two media
Transmitter a device that gives out a signal, such as sound in a sonar transmitter
Transducer a device that changes an electrical signal into light or sound, or changes light or sound into an electrical signal
Transfer (of energy) shifting energy from one place to another
Translucent describes objects that transmit light but diffuse (scatter) the light as it passes through
Transmitted light or other radiation passed through an object
Transparent describes objects that transmit light; you can see through transparent objects
Transverse describes a wave in which the vibrations are at right angles to the direction the wave moves
Turbine a component in a generator that turns when kinetic energy is transferred to it from steam, water, or wind
Turning effect a force causing an object to turn
Turning force the moment of a force

Ultrasound sound at a frequency greater than 20 000 Hz, beyond the range of human hearing
Umbra the area of total shadow behind an opaque object where no light has reached
Unbalanced forces describes forces on an object that are unequal
Universe everything that exists
Upright describes an image that is the right way up
Upthrust the force on an object in a liquid or gas that pushes it up
Uranium a metal used in nuclear power stations
Uranus a large outer planet made of gas, seventh from the Sun
Useful energy the energy that you want from a process

Vacuum a space in which there is no matter
Variable a quantity that can change, such as time, temperature, length, or mass. In an investigation you should change only one variable at a time to see what its effect is.
Venus a rocky inner planet, second from the Sun
Vibrate to move continuously and rapidly to and fro
Vibration motion to and fro of the parts of a liquid or solid
Virtual describes an image that cannot be focused onto a screen
Volt the unit of measurement of voltage, symbol V
Voltage a measure of the strength of a cell or battery used to send a current around a circuit, measured in volts
Voltmeter a device for measuring voltage

Wasted energy energy transferred to non-useful forms, often thermal energy transferred to the surroundings
Water resistance the force on an object moving through water that causes it to slow down (also known as drag)
Watts the unit of power, symbol W
Wave a variation that transfers energy
Wave energy/power using energy in waves to generate electricity
Wavelength the distance between two identical points on the wave, such as two adjacent peaks or two adjacent troughs
Weight the force of the Earth on an object due to its mass
White dwarf a small, very dense star, part of the life cycle of our Sun
Wind energy/power energy from wind that can be used to generate electricity
Wind farm a collection of wind turbines
Wind turbine a turbine and generator that uses the kinetic energy of the wind to generate electricity

Year the length of time it takes for a planet to orbit the Sun